# Essentials of Soft Matter Science

# Essentials of Soft Matter Science

Françoise Brochard-Wyart, Pierre Nassoy,
and Pierre-Henri Puech

CRC Press
Taylor & Francis Group
Boca Raton London New York

CRC Press is an imprint of the
Taylor & Francis Group, an **informa** business

Additional Exercises and Instructor's Manual are available for download at www.crcpress.com/9781498773928

CRC Press
Taylor & Francis Group
6000 Broken Sound Parkway NW, Suite 300
Boca Raton, FL 33487-2742

© 2020 by Taylor & Francis Group, LLC
CRC Press is an imprint of Taylor & Francis Group, an Informa business

No claim to original U.S. Government works

Printed on acid-free paper

International Standard Book Number-13: 978-1-138-74276-5 (Hardback)
International Standard Book Number-13: 978-1-4987-7392-8 (Paperback)

*First published in the French language by Dunod (© 2018) under the title 'Physique de la matière molle'*

**Visit the Taylor & Francis Web site at**
**http://www.taylorandfrancis.com**

**and the CRC Press Web site at**
**http://www.crcpress.com**

# Contents

# Acknowledgments

THIS BOOK IS WRITTEN in the spirit of Pierre-Gilles de Gennes. We tried to respect his teaching and research style. Remarks and suggestions from Madeleine Veyssié were very beneficial. We also wish to thank Erdem Karatekin for careful reading, and we acknowledge Olivier Sandre, Axel Buguin, Claude Redon, Liliane Léger, Fabien Bertillot, Kévin Alessandri, Aurélien Roux, Stéphane Douezan, and David L. Hu for sharing with us some unpublished photographs. Finally, we are grateful to Jean-Francois Joanny and L. Mahadevan for their encouragement to complete this book.

Parts of the present texts and images were adapted from the book *Physique de la matière molle*, Dunod Editions (in French). We would like to thank Dunod Editions for permitting us to use this material for the present publication.

# Acknowledgments

# Authors

**Françoise Brochard-Wyart, PhD,** is a theoretical physicist in soft matter physics and Professor at Sorbonne University and the Institut Curie, France. She studied at École Normale Supérieure de Cachan. After a PhD in Liquid Crystals under the supervision of Pierre-Gilles de Gennes, she studied polymer physics and wetting before moving to biophysics. She is a member of Institut Universitaire de France. She was awarded the Jean Ricard Prize from the French Physical Society in 1998.

**Pierre Nassoy** is an experimentalist physicist and a CNRS senior scientist at the Institut d'Optique d'Aquitaine in Bordeaux, France. He studied at Ecole Supérieure de Physique et Chimie Industrielles de Paris (ESPCI) and earned an engineering diploma. He was a junior CNRS scientist interested in cell biophysics at the Institut Curie until 2012.

**Pierre-Henri Puech** is an experimentalist (bio)physicist and an Inserm junior scientist at the Laboratoire Adhésion et Inflammation (LAI) in Marseille, France. He studied at Ecole Supérieure de Physique et Chimie Industrielles de Paris (ESPCI) and earned an engineering diploma. His research interests center around cell biophysics, in particular in the context of mechanotransduction in T-cell recognition.

# CHAPTER 1

# Introduction

## 1.1 THE BIRTH OF SOFT MATTER

P.-G. de Gennes (1932–2007) (Figure 1.1) is considered the inventor of the science called soft matter. After spectacular initial contributions in solid physics (magnetism, superconductivity), his career in the theoretical physics of condensed matter opened up to a very wide spectrum of subjects, namely liquid crystals, polymers, colloids, wetting and adhesion, and biophysics, which define soft matter. Although de Gennes' record is very impressive, the importance of his work mainly relies on his style, in permanent contact with the experimentalists and the industrial world, and on the idea that all the physical phenomena can be explained in simple terms. Pierre-Gilles had the passion to transmit his knowledge and discoveries to a wide audience, from schoolchildren to researchers, with accurate and colorful words and an impetus that triggered scientific vocations. His work was rewarded by the Nobel Prize in Physics in 1991.

We shall start this book on the physics of soft matter with some excerpts from the Nobel lecture delivered by Pierre-Gilles de Gennes in Stockholm in December 1991.

FIGURE 1.1   Pierre-Gilles de Gennes.

### 1.1.1 What Do We Mean by Soft Matter?

*Americans prefer to call it "complex fluids". This is a rather ugly name, which tends to discourage the young students. But it does indeed bring in two of the major features:*

> *Complexity. We may, in a certain primitive sense, say that modern biology has proceeded from studies on simple model systems (bacteria) to complex multicellular organisms (plants, invertebrates, vertebrates...). Similarly, from the explosion of atomic physics in the first half of this century, one of the outgrowths is soft matter, based on polymers, surfactants, liquid crystals, and also on colloidal grains.*

> *Flexibility. I like to explain this through one early polymer experiment, which has been initiated by the Indians of the Amazon basin: they collected the sap from the hevea tree, put it on their foot, let it "dry" for a short time. And, behold, they have a boot. From a microscopic point of view, the starting point is a set of independent, flexible polymer chains. The oxygen from the air builds in a few bridges between the chains, and this brings in a spectacular change: we shift from a liquid to a network structure which can resist tension – what we now call a rubber (in French: caoutchouc, a direct transcription of the Indian word). What is striking in this experiment, is the fact that a very mild chemical action has induced a drastic change in mechanical properties: a typical feature of soft matter.*

### 1.1.2 Style of Research in Soft Matter

#### 1.1.2.1 Simple Experiments

*I would like now to spend a few minutes thinking about the style of soft matter research. One first, major, feature, is the possibility of very simple experiments.(...) Let me take the example of surfactants, molecules with two parts: a polar head which likes water, and an aliphatic tail which hates water. Benjamin Franklin performed a beautiful experiment using surfactants; on a pond at Clapham Common, he poured a small amount of oleic acid, a natural surfactant which tends to form a dense film at the water-air interface. He measured the volume required to cover all the pond. Knowing the area, he then knew the height of the film, something like three nanometers in our current units. This was, to my knowledge, the first measurement of the size of molecules. In our days, when we are spoilt with exceedingly complex toys, such as nuclear reactors or synchrotron sources, I particularly like to describe experiments of this Franklin style to my students.*

*Let me quote two examples. The first concerns the wetting of fibers. Usually a fiber, after being dipped in a liquid, shows a string of droplets, and thus, for some time, people thought that most common fibers were non-wettable. F. Brochard analysed theoretically the equilibria on curved surfaces, and suggested that in many cases we should have a wetting film on the fiber, in between the droplets. J.M. di Meglio and D. Queré established the existence, and the thickness, of the film, in a very elegant way [1]. They created a pair of neighbouring droplets, one small and one large, and showed that the small one emptied slowly into the big one (as capillarity wants it to go). Measuring the speed of the process, they could go back to the thickness of the film which lies on the fiber and connects the two droplets: the Poiseuille flow rates in the film are very sensitive to thickness. Another elegant experiment in wetting concerns the collective modes of a contact line, the edge of a drop standing on a solid. If one distorts the line by some external means, it returns to its equilibrium shape with a relaxation rate dependent upon the wavelength of the distortion, which we wanted to study. But how could we distort the line? I thought of very complex tricks, using electric fields from an evaporated metal comb, or other, even worse, procedures. But Thierry Ondarcuhu came up with a simple method.*

1) *He first prepared the unperturbed contact line L by putting a large droplet on a solid.*

2) *He then dipped a fiber in the same liquid, pulled it out, and obtained, from the Rayleigh instability, a very periodic string of drops.*

3) *He laid the fiber on the solid, parallel to L, and generated a line of droplets on the solid.*

4) *He pushed the line L (by tilting the solid), up to the moment where L touched the droplets; then coalescence took place, and he had a single, wavy line on which he could measure relaxation rates* [2].

### 1.1.2.2 Theory

1.1.2.2.1 Reducing a Complex Problem to the Essence    Working with experimentalists or engineers in industry, and later with biologists, when confronted with situations involving a large number of parameters, P.-G. de Gennes had the art of unveiling the central physical phenomenon. We are used to characterizing this approach by comparing it with the style of Picasso, in particular the famous series of "Bull" lithographs, in which Picasso sketches a bull with less and less detail to finish with a few lines. These drawings profoundly marked Pierre-Gilles de Gennes (Figure 1.2):

*Everyone has his treasure of images of which we only had a glimpse but that we never forget. An example for me: Picasso painting with large white lines on a window and filmed by Clouzot. Everything I tried to painstakingly draw later was born from those moments.*

(Pierre-Gilles de Gennes in *L'émerveillement* [3])

FIGURE 1.2 The Abduction of Sabines, drawn by Pierre-Gilles de Gennes in 1983 (left) during a stay in Florence (right: the sculpture of Giambologna in the Loggia dei Lanzi). Private collection and picture from FBW.

1.1.2.2.2 Formulating a Problem by Using Dimensional Arguments and Scaling Laws    P.-G. de Gennes strives to give simple analytical expressions for his findings, even if they are most often the result of complex calculations. He always gives a physical interpretation and uses drawings to explain it, like the image of the blobs or the reptation model, which opened up the physics of polymers to a wide audience.

1.1.2.2.3 Having a Broad Scientific Culture to Make Analogies between Various Disciplines *I have emphasized experiments more than theory. Of course, we need some theory when thinking of soft matter. And in fact, some amusing theoretical analogies sometimes show up between soft matter and other fields. One major example is due to S.F. Edwards [4]. Edwards showed a beautiful correspondence between the conformations of a flexible chain and the trajectories of a non-relativistic particle; the statistical weight of the chain corresponding to the propagator of the particle. In the presence of external potentials, both systems are ruled by exactly the same Schrödinger equation! This observation has been the key to all later developments in polymer statistics. Another amusing analogy relates the smectics A to superconductors. It was discovered simultaneously by the late W. McMillan (a great scientist, who we all miss) and by us. Later, it has been exploited artistically by T. Lubensky and his colleagues [5]. Here again, we see a new form of matter being invented. We knew that type II superconductors let in the magnetic field in the form of quantized vortices. The analog here is a smectic A inside which we add chiral solutes, which play the role of the field. In some favorable cases, as predicted in 1988 by Lubensky, this may generate a smectic phase drilled by screw dislocations – the so-called A\* phase. This was discovered experimentally only one year later by Pindak and coworkers [6], a beautiful feat.*

One could cite many other examples such as the analysis of the deformations of "closed" flexible membranes (red blood cell, vesicle, etc.) modeled by P.-G. de Gennes using the spherical model of phase transitions in two dimensions [7].

### 1.1.2.3 Doing Science and Having Fun

Let us close this introductory section with the words chosen by P.-G. de Gennes to close his Nobel lecture. These are four verses of a poem by François Boucher ("*La souffleuse de savon*" – the soap-blowing lady):

> *Amusons nous sur la terre et sur l'onde*
> *Malheureux qui se fait un nom*
> *Richesses, honneurs, faux éclats de ce monde*
> *Tout n'est que boules de savon*

> Have fun on sea and land
> Unhappy it is to become famous
> Riches, honors, false glitters of this world
> All is but soap bubbles

### References

1. J.M. di Meglio, *CR. Acad. Sci.*, 303 II, 437 (1986).
2. T. Ondarcuhu, M. Veyssié, *Nature*, 352, 418 (1991).
3. T. de Wurstemberg. *L'émerveillement.* Saint-Augustin Edition (1998).
4. S.F. Edwards, *Proc. Phys. Sot.*, 85, 613 (1965).
5. S.R. Renn, T. Lubensky, *Phys. Rev. A*, 38, 2132 (1988).
6. J.W. Goodby, M.A. Waugh, S.M. Stein, E. Chin, R. Pindak, J.S. Patel, *J. Am. Chem. Soc.*, 111, 8119 (1989).
7. M.N. Barber, M.E. Fisher, *Ann. Phys.*, 77, 1–78 (1973).

## 1.2 OVERVIEW

In this book we discuss the physics of soft matter. We believe that this book will complement previous ones on the same field of soft condensed matter physics. Each author or group of authors may have a specific view about how this discipline should be taught. We have deliberately attempted to report Pierre-Gilles de Gennes' vision on the discipline that he invented as faithfully as possible. The present book evolved from notes prepared for seminars, conferences, and lectures to undergraduate and graduate physics students by Pierre-Gilles de Gennes and one of us (FBW).

Following a general introduction that sets the main definitions and summarizes the basic knowledge about relevant physical interactions (Chapter 2) and phase transitions (Chapter 3), we cover, in the first part of the book, the main classical sub-fields of soft matter, i.e. interfaces (including colloids, wetting, dewetting – Chapter 4), liquid crystals (Chapter 5), surfactants (Chapter 6), and polymers (Chapter 7). In all these chapters, not only do we minimize the mathematical details, but we favor orders of magnitude, back-of-an-envelope calculations and scaling law arguments. In the second part, we aim at concretely exemplifying the previously introduced concepts with specific cases extracted from our everyday life (Chapter 8), technology (Chapter 9), and biology (Chapter 10). We address and decipher some remarkable processes or achievements such as the fabrication of flavor pearls in molecular cuisine, the magic of painting, and the collective behavior of fire ant swarms.

# CHAPTER 2

# Soft Matter

## 2.1 MESOSCOPIC COMPLEX SYSTEMS

Soft matter is a class of materials that includes polymers, liquid crystals, and detergents, which are also called complex systems or fluids. Their common feature is the presence of a mesoscopic scale that governs most properties of the system.

### 2.1.1 Mesoscale

The term mesoscale refers to an intermediate length scale between macroscopic objects and those of atomic size. The range between a few Angstroms (Å) and a few thousand Å is the typical length scale for most families of soft matter systems that we will study.

Figure 2.1 shows some examples of such systems. The fact that a soap film contains several hundred water molecules piled along the film thickness or that a chain of polymers is composed of more than a thousand monomeric units ensures that continuous theories remain valid to describe these systems. Since the number of particles $N$ is large ($N > 100$), statistical mechanics can be applied.

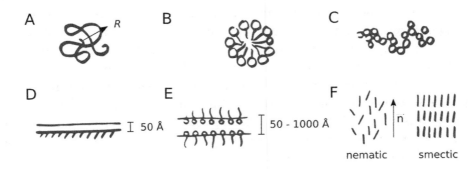

FIGURE 2.1 Examples of mesoscopic systems in soft matter. (A) Polymer chain ($N = 1000$; $R = N^\nu a$); (B) Surfactant micelle ($N = 100$); (C) Flocculation aggregate ($N = 500$); (D) Wetting film; (E) Soap film; (F) Liquid crystal.

## 2.1.2 Disorder

Disorder is an important feature of soft matter. Let us exemplify this property with two cases:

- Like crystals, liquid crystals have an orientational order. However, by contrast with crystals, the centers of gravity of the liquid crystal molecules present a translational disorder and they flow like liquids.

- Polymers are often compared to a dish of entangled spaghetti [1]. Due to thermal agitation, the long chains sneak by reptation among the others (Figure 2.2), like snakes in the savannah.

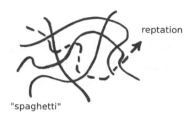

FIGURE 2.2   Reptation of a polymer chain.

## 2.1.3 Topology and Geometry

### 2.1.3.1 Connectivity

When water is dispersed in oil, an emulsion of water in oil is formed. Increasing the amount of water leads to the opposite configuration, namely an emulsion of oil in water. The transition between these two states corresponds to a percolation transition [2]. This is a threshold phenomenon.

Percolation is a concept that applies to a wide variety of systems. It occurs when you prepare your coffee. The first drop of coffee coming out of the brewer traces a continuous path of water through the coffee powder. In a different field, archipelagos of islands are separated from each other at high tide, but when the tide goes down, they become peninsulas and allow the traveler to explore them on foot. The spread of infectious diseases (e.g. influenza epidemics) or forest fires also illustrates this percolation phenomenon.

To return to complex systems, the process of polymer vulcanization that occurs when liquid latex turns into a solid (rubber) by bridging the polymer chains (Figure 2.3) is characterized by a percolation transition: when a critical number of crosslinking points (nodes) is reached, the system that was liquid becomes a solid.

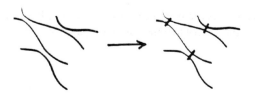

FIGURE 2.3   Vulcanization of polymers: from latex (liquid) to rubber (solid).

### 2.1.3.2 Self-Similarity

While some systems such as latex particles may be characterized by a single characteristic size, other systems are multi-scale or fractal [3].

In general, fractal structures are self-similar: the architecture of the object remains identical when one zooms in. Examples of these structures are the Romanesco cabbage, the snowflake, or the Brittany coast (Figure 2.4).

FIGURE 2.4   Examples of fractals. (A) Romanesco cabbage; (B) Snowflakes; (C) Coast of Brittany. (Copyright Shutterstock.)

A fractal system is characterized by the fractal dimension $D_f$. Let us take the example of a fractal line. Its length $L$ depends on the scale $\varepsilon$, over which the measure is carried out. Thus, the measure of the Brittany coast with a scale unit $\varepsilon = 1$ km or with $\varepsilon = 1$ m will not give the same result. The relationship $L(\varepsilon)$ defines $D_f$:

$$L(\varepsilon) = \left(\frac{L}{\varepsilon}\right)^{D_f} \varepsilon,$$

where $L$ is the overall length. A similar definition can be applied to fractal surfaces (Table 2.1). The fractal dimension $D_f$ is less than or equal to d, the dimension of the space.

TABLE 2.1   Fractal Dimension $D_f$

|  | Dots | Smooth Line | Smooth Surface | Fractal Line | Fractal Surface |
|---|---|---|---|---|---|
| $D_f$ | 0 | 1 | 2 | $1 < D_f < 3$ | $2 < D_f < 3$ |
|  |  | $D_f \in \mathbb{N}$ |  | $D_f \in \mathbb{Z}$ |  |

The Koch curve or snowflake is a classic mathematical example of a fractal line (Figure 2.5). By construction, we see that if $\varepsilon$ is divided by 3, the length is multiplied by 4/3, which results in:

$$L\left(\frac{\varepsilon}{3}\right) = \left(\frac{4}{3}\right) L(\varepsilon)$$

Using the equation that defines $D_f$, we find that the fractal dimension is:

$$D_f = \frac{\ln 4}{\ln 3}$$

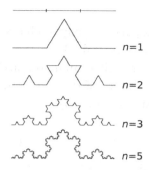

FIGURE 2.5 Construction of the Koch flake.

In the case of a linear polymer, the chain of $N$ monomers is cut into units called "blobs" (Figure 2.6). Each blob has a size $\varepsilon$ and contains $g$ monomers.

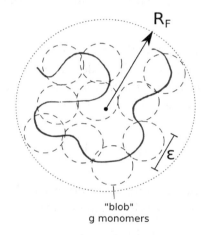

FIGURE 2.6 Polymer chain split into "blob" sub-units.

In Section 7.3, we will define the exponent $\nu$, which characterizes the conformation of the polymer chain, and we will show that the relationship between size and number of monomers is valid at all scales:

$R = N^{\nu}$, $\varepsilon = g^{\nu}a$. This relationship allows us to write:

$$L(\varepsilon) = \left(\frac{N}{g}\right)\varepsilon = \left(\frac{R}{\varepsilon}\right)^{1/\nu}\varepsilon,$$

and we obtain the fractal dimension of the polymer chain (Table 2.2).

$$D_f = \frac{1}{\nu}.$$

Table 2.2 gives the $D_f$ values for different chain conformations.

TABLE 2.2   Fractal Dimension $D_f$
of a Polymer Chain

|  | Ideal Chain | Collapsed Chain | Swollen Chain |
|---|---|---|---|
| $v$ | 1/2 | 1/3 | 3/5 |
| $D_f$ | 2 | 3 | 5/3 |

**Practical Question:** What does the fractal dimension of a metro network tell us about the growth of a city?

Fractals were introduced in the 1970s by the mathematician Benoît Mandelbrot. We can construct fractal objects by using mathematical methods (e.g. the Koch curve – Figure 2.5), but we also find physical fractal objects that form "naturally" like ferns, seacoasts, or networks of blood vessels and neurons. It is sometimes just a question of curiosity to wonder whether a structure, which attracts our attention because its irregular or ramified nature seems self-similar, is truly fractal. Then the question will arise about the meaning of the calculated fractal dimension.

This approach was taken by L. Benguigui and M. Daoud [4] when they asked whether the metro and regional express (RER) networks in Paris were fractal. This network developed from the beginning of the 20th century, starting with six *intra muros* lines, then extending to the suburbs. The advantage of studying the Parisian network is that the city of Paris is relatively circular, which simplifies the analysis. Figure 2.7A shows a schematic version of the lines, where the dots represent the stations. By counting the number of $N(R)$ stations located in circles of increasing radius $R$, we may anticipate two extreme cases. If the city is uniformly covered with stations, their density, $\rho(R) = N(R)/4\pi R^2$, is constant, hence $N(R) \sim R^2$.

Conversely, if stations are equidistant from each other along radially oriented lines, we may expect $N(R) \sim f.R/d \sim R$. For a random distribution, we have $N(R) \sim R^{D_f}$, which defines the fractal dimension $D_f$.

The outcome of the study is shown in Figure 2.7B in logarithmic scale. There are two regimes. For distances $R < R_0 = 6.5$ km, $N(R) \sim R^2$. For $R > R_0$, $N(R) \sim R^{0.47}$, i.e. $D_f \sim 0.5$. The city of Paris limited by the ring road (i.e. excluding the suburb) has precisely a radius of 6.5 km. While the intramural subway network has a fractal dimension of two, meaning that it is compact, it becomes fractal beyond the ring road and even very sparse (because $D_f < 1$), which means that the distant suburban cities are still underserved.

Similarly, the authors have shown that the total length of the RER lines that connect the center of Paris to the suburbs, $L(R)$, also has a fractal dimension, $D_f = 1.47$ (Figure 2.7C).

By focusing on the subway network, Paris, Moscow, and Berlin are all characterized by a fractal dimension $D_f = 1.7$–$1.8$ [5], which corresponds to an optimal value. Coincidentally or not, 1.7 is also the fractal dimension of aggregates of colloidal particles whose formation is limited by diffusion [6].

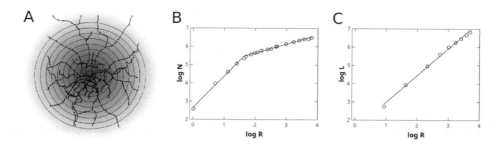

**FIGURE 2.7** (A) Drawing of the Parisian rail network (Metro and RER). (B) Plot of the number of stations $N$ as a function of the distance $R$ to the center of the city. (C) Plot of the total length $L$ of the lines as a function of the distance $R$ to the center of the city.

Further, one can also use the fractal dimension of a transportation network to represent the growth of a city. For example, in Seoul, between 1980, which marked the construction of the first subway lines, and the early 2000s, the fractal dimension of the network has increased from 1.15 to 1.35 [7]. Clearly, the Seoul metro network has not yet reached saturation.

## References

1 P.-G. de Gennes, Jacques Badoz, *Fragile Objects*. Springer, 1996.
2. P.-G. de Gennes, *La Recherche*. Sophia Publications, 1976.
3. A. Lesnes, M. Laguës, *Scale Invariance: From Phase Transitions to Turbulence*. Springer, 2012.
4. L. Benguigui, M. Daoud, *Geogr. Anal.*, 23, 362–368 (1991).
5. L. Benguigui, *Environ. Plan. A*, 27, 1147–1161 (1995).
6. T.A. Witten, L.M. Sander, *Phys. Rev. Lett.*, 47, 1400 (1981).
7. K.S. Kim, L. Benguigui, M. Marinov, *Cities*, 20, 31–39 (2003).

## 2.2 FRAGILE OBJECTS

Soft matter objects or complex fluids form fragile structures, which can be formed or broken apart easily, because the interactions are weak, of the order of the thermal energy.

### 2.2.1 Weak Interactions

Solid state physics is the field of strong bonds (ionic, covalent, electronic bonds) whose energies are of the order of the electron-volt (1 $eV = 1.6 \times 10^{-19}$ Joule).

Soft matter is the field of weak bonds, characterized by interactions of the order of the thermal agitation $k_B T$ (where $k_B$ is the Boltzmann constant and $T$ is the temperature). At room temperature ($T = 300$ K), $k_B T = 1/40\ eV = 4\ 10^{-21}$ Joule.

The four main types of weak interactions are:

- *Van der Waals interactions* (Figure 2.8A). They are ubiquitous, long-range, and generally attractive. For example, grains interacting through van der Waals interactions have a binding energy $U \approx (R/d) k_B T$, with $R$ the radius of the particle and $d$ the distance to contact. For particles of radius $R$ of the order of 1 μm and $d \sim 1$ nm, $U$ is of the order of 100 $k_B T$. Thermal agitation is not sufficient to separate them.

- *Electrostatic interactions.* They exist in aqueous medium between charged particles. Their range is controlled by the concentration of salt and the valence of the counter-ion. They are repulsive between particles of the same nature. The cloud of counter-ions shown in Figure 2.8B prevents the particles from coming in contact and sticking together due to van der Waals interactions.

- *Steric interactions.* They are important in the case of polymer-based systems or surfactant assemblies. They are long-range and repulsive. The *corona* shown in Figure 2.8C is the "hairy" cloud of monomers that prevents the particles from getting closer.

- *Solvation and hydrogen bonding interactions.* They result from the specific structure of some liquids (particularly water). They are very short range and directional (Figure 2.8D).

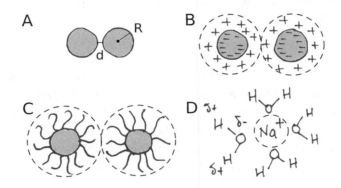

FIGURE 2.8  (A) Van der Waals interaction. (B) Electrostatic stabilization. (C) Steric stabilization. (D) Solvation.

## 2.2.2 Large Responses

In complex systems, small perturbations may produce great effects.

- *Liquid crystals* (Chapter 5)

  They are sensitive to electric ($\vec{E}$) and magnetic ($\vec{B}$) fields. Liquid crystal displays work through the action of very weak electric fields that may switch the orientation of LC molecules and thus modulate the displayed image.

- *Foams* (Section 6.3)

  Traces of detergent (or surfactant) dispersed in water generate a foam after shaking.

- *Sol-gel transition* (Chapter 3)

  Traces of oxygen in the air transform the sap of the Hevea tree into solid rubber. But oxygen is too reactive and ends up cutting the molecules. Rubber boots made with Hevea rubber by the Indians (Figure 2.9) did not last very long: they broke up and eventually moldered on their feet. In 1830, Goodyear showed that rubbers remained stable for years by replacing oxygen with sulfur. This finding led to the fabrication of tires that equip all our cars.

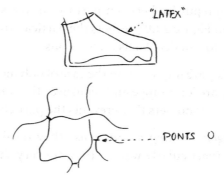

FIGURE 2.9 The Indian experiment drawn by P.-G. de Gennes: Top: wrapped foot; Bottom: solid with oxygen crosslinks ("ponts" in French).

- *Colloidal flocculation* (Section 4.1)

  A small amount of salt triggers the flocculation of a colloidal suspension. This process is especially used for the treatment of wastewater. In 1800, Faraday performed experiments with gold colloids (Figure 2.10). He noticed that the red coloration of the suspension turned to bluish upon flocculation induced by salinity (or ionic strength) increase of the solution. Conversely, by adding gelatin, the solution remained stable, highlighting the stabilizing effect due to steric interactions.

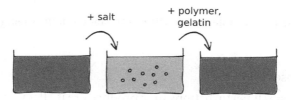

FIGURE 2.10 The Faraday experiment: Flocculation of gold colloids induced by salt and stabilization by addition of polymers.

- *Surfactant films* (Section 6.2)

  In ancient times, sailors calmed the stormy sea by throwing oil onto the waves. No more than a molecular monolayer on the surface of water lowers the surface tension of water (Chapter 4), alters the interfacial hydrodynamics, and causes a drop in the magnitude of the waves (see Benjamin Franklin's experiment summarized by P.-G. de Gennes in his Nobel lecture and sketched in Figure 2.11). But surface tension decrease has other consequences: duck feathers are no longer "water-proof," and insects like water spiders can no longer walk on water (Section 8.3).

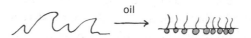

FIGURE 2.11 Change of interfacial hydrodynamics with a surfactant monolayer: from a rough sea to a waveless sea.

As P.-G. de Gennes wrote in the chapter "Soft matter: birth and growth of concepts" extracted from [1]:

> *If a condensed matter physicist is asked to imagine a system with strong responses to some perturbation, his first reaction will often involve critical phenomena: for instance, near the Curie point $T_c$ of a ferromagnet, the magnetic susceptibility $\chi$ is very large.*

Let us also mention liquid–gas transitions and demixing in binary mixtures as other well-documented examples.

This property is found in soft matter because these systems that are assembled by tenuous forces have giant fluctuations, similar to critical phenomena. For this reason, we have dedicated a full chapter to phase transitions and critical phenomena (Chapter 3). For example, in stretching a polymer chain (Section 7.2), the length of the chain, $L$, is proportional to the magnitude of the end-to-end distance fluctuations $\langle \vec{R}^2 \rangle$ (Figure 2.12).

$$\langle \vec{R} \rangle = \vec{0}, \ \langle \vec{R}^2 \rangle^{1/2} = R_F, \ L \sim \langle \vec{R}^2 \rangle f$$

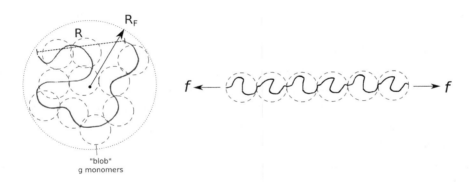

FIGURE 2.12  Stretching of a polymer chain by application of forces f at both ends.

### 2.2.3 Soft Matter and Biology

The concepts of soft matter also apply to biology. All biological architectures, including DNA molecules which are the main material of the genetic code, proteins and cell membranes, are governed by weak interactions, and they are in perpetual renewal. There would be no life without soft matter!

Let us mention for example that the separation of two strands of DNA by sequential rupture of hydrogen bonds between base pairs is achieved with very weak forces, of the order of the picoNewton ($f \sim 10^{-12}$ N) (Figure 2.13A). Similarly, a red blood cell can sneak into micro-vessels, revealing its exceptional elasticity (small stress, large deformation) (Figure 2.13B).

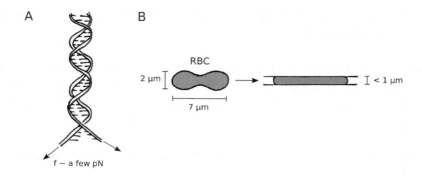

FIGURE 2.13 (A) Opening of two DNA strands. (B) Squeezing of a red blood cell (RBC) into a micro-vessel.

## Reference

1. L.M. Brown, A. Pais, Sir B. Pippard, eds., *Twentieth Century Physics*. IOP Publishing Ltd, AIP Press Inc., 1995, vol. III, pp. 1593–1616.

## 2.3 VAN DER WAALS FORCES

Van der Waals forces are ubiquitous and generally attractive. Their origin is dipolar.

### 2.3.1 Classification and Range of van der Waals Interactions

Van der Waals forces, devised by Johannes Diderick van der Waals (1837–1923), have three physical origins. They may take place between polar molecules (Keesom), between polar and non-polar molecules (Debye), and between non-polar molecules (London). They all lead to an attractive interaction that decays with distance $r$ as $1/r^6$.

Let us take here the case of non-polar molecules. Fluctuations in the electron cloud of a molecule A create an instantaneous dipole $\vec{\mu}_i$. The molecule B is polarized under the effect of the electric field $\vec{E}_B$ created by $\vec{\mu}_i$ ($\langle \mu \rangle = 0$):

$$E_B \approx \frac{\mu_i}{r_{AB}^3} \frac{1}{4\pi\varepsilon_0},$$

where $r_{AB}$ is the distance between A and B (Figure 2.14).

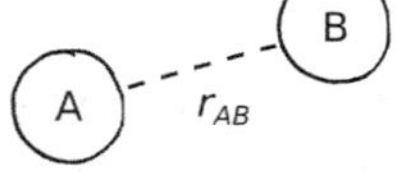

FIGURE 2.14 Instant dipole $\mu_i$ borne by molecule A creates an induced dipole $\mu_B$ on molecule B. A and B are distant from $r_{AB}$.

The molecule B thus acquires an induced dipole $\mu_B$ defined by:

$$\mu_B = \alpha_B E_B,$$

where $\alpha_B$ is the polarizability coefficient of A.

The interaction energy $\varepsilon$ between A and B is given by:

$$\varepsilon = -\frac{1}{2}\mu_B E_B = -\frac{1}{2}\alpha_B \left\langle E_B^2 \right\rangle.$$

However, the fluctuations of $\vec{E}_B$ are proportional to $\alpha_A$, the polarizability coefficient of the molecule A. We thus obtain:

$$\varepsilon = -k\frac{\alpha_A \alpha_B}{r^6}$$

### 2.3.2 Van der Waals Interactions between Two Media

Knowing the molecular origin and the expression of the van der Waals energy between two molecules, we now want to obtain an expression for this interaction between two media, *1* and *2*. The density of molecules in the medium $i$ ($i = 1, 2$) is noted $\rho_i$. The polarizability per unit of volume of medium $i$ is $\rho_i \alpha_i$.

#### 2.3.2.1 Hamaker Constant

The Hamaker constant is defined by:

$$A = \pi^2 C \rho_1 \rho_2,$$

where $C$ is the van der Waals potential coefficient discussed in the previous paragraph: $C = k\alpha_1\alpha_2$.

An order of magnitude calculation shows that $A \approx k_B T$. The Hamaker constant will be used to describe the interaction between two media by summing up the interactions between all molecules (Figure 2.15).

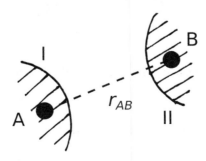

FIGURE 2.15   Interaction between two media I and II in vacuum.

### 2.3.2.2 Interactions between Two Half-Spaces Separated by Vacuum

Let us start by calculating the interaction energy U between a molecule and an infinite half-space (Figure 2.16). This consists of summing all pair interactions. Considering a ring of width $dr$, thickness $dz$, and radius $r$ in the medium 2, the energy U is written as:

$$U = -C \int_h^\infty dz \int_0^\infty \rho_2 \frac{2\pi r dr}{(r^2 + z^2)^3} = -C\rho_2 \int_h^\infty \frac{\pi}{2} \frac{dz}{z^4} = -C \frac{\pi \rho_2}{6h^3}$$

The energy density (energy per unit area) between two half-spaces, 1 and 2, separated in a vacuum by a distance D, is obtained by integrating over the variable $z$ from D to infinity in the medium 1.

$$E = -C\rho_1\rho_2 \frac{\pi}{6} \int_D^\infty \frac{dz}{z^3} = -C \frac{\pi}{12} \rho_1\rho_2 \frac{1}{D^2}$$

By definition of the Hamaker constant, E reads:

$$E = -\frac{A_{12}}{12\pi} \frac{1}{D^2},$$

which shows that the van der Waals interaction between two plates is long-range and goes as $1/(\text{distance})^2$ instead of $1/(\text{distance})^6$.

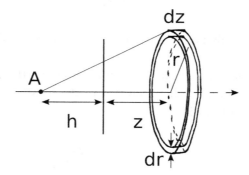

FIGURE 2.16 Notations for the calculation of the van der Waals interaction between a molecule and a semi-infinite medium.

### 2.3.2.3 Interactions between Two Plates Separated by a Dielectric Medium

This configuration corresponds for instance to the practical situation in which a liquid (L) film of thickness e is deposited on a solid (S) surface, in contact with the vapor (V) (Figure 2.17). Since interfacial tensions are defined for semi-infinite media, there are correction terms for films of microscopic thickness that we are going to detail here.

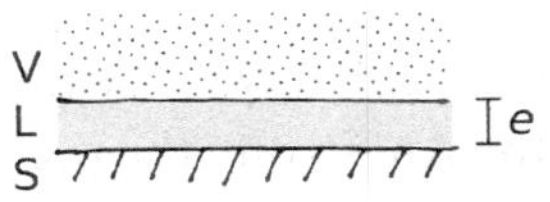

FIGURE 2.17   Interactions between two media (S, V) separated by a dielectric material (L).

We use the notations defined in Chapter 4.

- In the limit $e = 0$, the energy per unit area of the system is:

$$E = \gamma_{so} \text{ (surface energy of the bare solid).}$$

- In the limit of a thick film, $e \to \infty$, the surface energy of the system is the sum of the surface energy of the solid in contact with the liquid $\gamma_{SL}$ and that of the liquid with the vapor $\gamma$:

$$E = \gamma_{SL} + \gamma$$

- In the general case, where the thickness e of the liquid film has a mesoscopic size, we set:

$$E = \gamma_{SL} + \gamma + P(e),$$

where *P(e)* is the corrective term due to long-range forces and obeys the boundary conditions $P(\infty) = 0$ and $P(0) = \gamma_{so} - (\gamma_{SL} + \gamma) = S$ (where $S$ will be defined as the spreading parameter). $P(e)$ represents the contribution of long-range interactions and must cancel if L and V or L and S are identical. We can therefore express $P(e)$ in a quadratic form with all polarizabilities: $P(e) = -k(e) \cdot (\alpha_L - \alpha_S)(\alpha_L - \alpha_V)$, where $k(e)$ is a constant for a given thickness $e$.

Now, if $S$ and $V$ are in vacuum, we know that: $P(e) = -\left(A_{LL}/12\pi\right)\left(1/e^2\right)$. It therefore leads to:

$$P(e) = \frac{A}{12\pi} \frac{1}{e^2}$$

where $A$ is the effective Hamaker constant defined by:

$$A = -\left(A_{LL} - A_{LS} + A_{SV} - A_{LV}\right)$$

In the particular case of a wetting film, the medium $V$ is air, which has a negligible density compared to that of the liquid $L$ and the solid $S$. By considering $A_{SV} = A_{LV} \approx 0$, the previous expression becomes:

$$P(e) = \frac{A_{LS} - A_{LL}}{12\pi e^2} = \frac{A}{12\pi e^2}$$

If $A > 0$, i.e. if the solid is more polarizable than the liquid, a liquid film (of thickness < 100 nm) will be stable. In the opposite case, $A < 0$, the film will be unstable.

The first derivative of $P(e)$ defines the disjunction pressure $\Pi$:

$$\Pi = -\frac{dP(e)}{de} = \frac{A}{6\pi e^3}$$

By introducing a molecular length: $a = \left(A/6\pi\gamma\right)^{1/2}$ (with $\gamma$ the surface tension of the liquid), which is typically of the order of the Angström, the previous relation becomes:

$$\Pi = \frac{\gamma a^2}{e^3}$$

$\Pi$ represents the decrease of the chemical potential of the liquid in the film. It is thus possible to calculate the thickness $e$ of a film which condenses at a height $H$ above the liquid reservoir ($\Pi(e) = \rho g H$).

### 2.3.2.4 Interaction between Two Identical Spheres

A calculation similar to the one described above shows that the van der Waals interaction energy between two spheres of radius $R$ separated by a distance $D$ is:

$$U = -\frac{A}{12}\frac{R}{D}.$$

The interaction force $F_{vdW}(D) = -\left(\partial U/\partial D\right)$ reads:

$$F_{vdW}(D) = -\frac{A}{12}\frac{R}{D^2}.$$

For two colloidal particles of radius $R = 1$ μm at a distance $D = 1$ nm apart ($R/D \approx 100$), we immediately remark that $U \gg k_B T$, and $F_{vdW}$ is of the order of 100 pN, i.e. more than 1,000 times the weight of a particle.

### 2.3.2.5 Experimental Validation with the Surface Force Apparatus (J. Israëlachvili)

The first instrument that allowed the directly measurement of the van der Waals interaction forces was the surface force apparatus designed by Jacob Israëlachvili et al. [1] (Section 2.5). The principle consists of detecting the attraction exerted between two curved and atomically smooth (mica) surfaces. One of the two surfaces is connected to a spring (Figure 2.18), and the other is displaced over a range of inter-distances from an Angstrom to hundreds of nanometers.

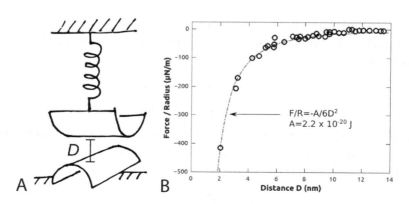

FIGURE 2.18 Israëlachvili Surface Force Apparatus. (A) Drawing of the crossed half-cylinders and the spring that enables force measurement as a function of distance $D$. (B) Typical curve Force/Radius vs. Distance. (Adapted from [1].)

*2.3.2.6 Two Identical Media Always Attract Each Other*

If one sets $S \equiv V \equiv 1$ and $L \equiv 2$ in the general expression of $P(e)$, we obtain:

$$P(e) = -\frac{\left(A_{11} + A_{22} - 2A_{12}\right)}{12\pi e^2} = -k\frac{\left(\alpha_1 - \alpha_2\right)}{12\pi e^2} < 0$$

As an application, two polymers A and B are immiscible if there are only van der Waals forces.

Let $u$ be the monomer–monomer interaction energy. For one chain containing N monomers (Figure 2.19), the interaction energy is larger than thermal energy:

$$\frac{Nu}{k_B T} \gg 1,$$

which leads to a phase separation between the two polymers.

FIGURE 2.19    Two polymers A and B are immiscible if the energy u between monomers is attractive (Van der Waals type). u is the interaction between a monomer A and a monomer B.

Reference

1. J. Israëlachvili, *Intermolecular and Surface Forces*. Academic Press, 1985.

## 2.4 ELECTROSTATIC INTERACTIONS

The electrostatic interaction between two charges $q_1$ and $q_2$ placed in a medium of dielectric constant $\varepsilon$ and distant from $r$ is given by Coulomb's law:

$$V(r) = \frac{q_1 q_2}{4\pi\varepsilon}\frac{1}{r}$$

The repulsion between two point charges is therefore long-range (in $1/r$).

Here, we will consider the interaction between two charged surfaces immersed in an aqueous solution.

### 2.4.1 Origin of the Surface Charge

In aqueous solution, two main mechanisms allow the creation of surface charges:

- *Surface groups are ionized.* The molecules located at the surface undergo a modification that leads to the dissociation of charged groups of atoms, leaving a charged surface. This is the case for glass or silica. In volume, silica is made of $SiO_2$ groups. But on the surface, there are silanol groups (–Si–OH). In contact with acidic or neutral water, these groups lose a proton and become negatively charged (–Si–O⁻) (Figure 2.20).

- *Charged ions or species present in solution bind to the surface.* Polyelectrolytes or ionic surfactants (such as SDS, sodium dodecyl sulfate) can adsorb onto glass. Small ions ($Ca^{2+}$) can associate with the polar heads of amphiphilic molecules or lipids.

$$SiOH \rightleftharpoons SiO^- + H^+ \qquad \text{in acid medium}$$
$$SiOH + H_2O \rightleftharpoons SiOH_2^+ + OH^- \qquad \text{in basic medium}$$

FIGURE 2.20   Ionization of silanol groups on the glass surface.

## 2.4.2 Electrostatic Double Layer

Let us consider a negatively charged surface. These charges create an electrical surface potential. In water (with or without added electrolyte), the counterions (here, positive charges) are attracted to the negative surface. At equilibrium, the surface charge is compensated by nearby counterions. But these counterions move under thermal agitation; they do not stick to the surface: they form a diffuse zone. The thickness of the cloud of counter ions in water in the presence of salt is characterized by a length, $\kappa^{-1}$, the Debye length, which defines the range of electrostatic interactions and which strongly depends on salt concentration, $c$.

The relation between the electrostatic potential $\Psi$ and the charge density $\rho$ is given by Poisson's law:

$$\Delta \Psi = -\frac{\rho}{\varepsilon},$$

where $\varepsilon$ is the dielectric permittivity of the medium. For example, for water, $\varepsilon = 80\, \varepsilon_0$, with $\varepsilon_0$ the vacuum dielectric constant ($\varepsilon_0 = 8.85 \times 10^{-12}$ F.m⁻¹).

The total charge density $\rho$ is the sum of the densities of the two ionic species. With $n_+$ (resp. $n_-$) the number of counter-ions (or co-ions) per unit volume, $z_+$ (resp. $z_-$) the valence of the counter-ions (respectively co-ions) and $e$ the charge of the electron, we have:

$$\rho = n_+ z_+ e - n_- z_- e$$

In addition, each ion distribution obeys a Boltzmann distribution:

$$n_+ = n_+ (\infty) \exp\left( -\frac{z_+ e \Psi}{k_B T} \right)$$

$$n_- = n_- (\infty) \exp\left( +\frac{z_- e \Psi}{k_B T} \right)$$

For instance, if the added salt is monovalent, $n_-(\infty) = n_+(\infty) = n_0$ is the density (in number) of ions in solution (Figure 2.21).

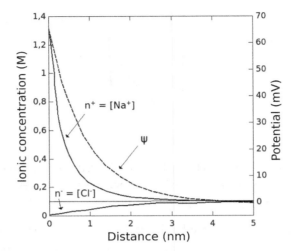

FIGURE 2.21 Distribution of counter-ions (solid lines) near a negatively charged surface and electrostatic potential $\Psi$ (dotted line). (Adapted from [1].)

The equation giving $\Psi(x)$ is the Poisson–Boltzmann equation. It is nonlinear in its general form. In the limit of low surface potentials (typically < 25 mV), it can be linearized in the form: $d^2\Psi / dx^2 = \kappa^{-2}\Psi$, with $\kappa^2 = \Sigma n_i z_i^2 e^2 / \varepsilon k_B T$.

The solution $\Psi$ that must satisfy both boundary conditions, $\Psi(\infty)=0$ et $\Psi(0) = \Psi_S$ is $\Psi(x) = \Psi_S \exp(-\kappa x)$. The surface potential $\Psi_S$ is related to the surface charge by Gauss's theorem, $E_S = -d\Psi/dx(x=0) = \sigma/\varepsilon$, which gives: $\kappa\Psi_S = \sigma/\varepsilon$

Interactions are screened over the Debye–Hückel length $\kappa^{-1}$.

When the charge density is set by the concentration $c$ ($c = n_0/N_A$, with $N_A = 6.02 \times 10^{23}$ the Avogadro number) of the added electrolyte (e.g. NaCl), $\kappa^{-1}$ is directly proportional to $c^{-1/2}$. The range of electrostatic interactions thus decreases with increasing salt concentration (see Table 2.3).

TABLE 2.3  Screening Length Values for a Monovalent Salt ($z=1$) as a Function of Concentration

| $c$ (mol.l$^{-1}$) | $10^{-1}$ | $10^{-3}$ | $10^{-5}$ |
|---|---|---|---|
| $\kappa^{-1}$ (Å) | 10 | 100 | 1,000 |

### 2.4.3 Repulsion between Two Charged Plates

When two charged surfaces are separated from each other by a distance $D \gg \kappa^{-1}$, they behave as neutral surfaces, and the electrostatic interaction is generally negligible when compared to van der Waals interaction. On the other hand, if $D < \kappa^{-1}$, there is overlap and contraction of the counter-ion clouds (Figure 2.22). Both plates repel each other.

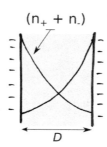

FIGURE 2.22 Repulsion between two charged surfaces loaded with the same negative surface density. The excess of counter-ions increases the ionic concentration between the plates if $\kappa D < 1$ and leads to a strong osmotic pressure which pushes the plates away.

We show that the pressure $P(D)$ felt by the plates is directly related to the increase of the ion concentration in the median plane (i.e. to the excess of osmotic pressure). The resolution of the Poisson–Boltzmann equation leads to:

$$P(D) = 64 n_0 k_B T \gamma_0^2 \exp(-\kappa D),$$

where $\gamma_0 = \tanh\left(e\Psi_S / 4k_B T\right)$ is defined from the electrostatic potential $\Psi_S$ at the surface of the plate.

An electrostatic free energy $U_{el}(D)$ is associated to this pressure and is defined by:

$$U_{el}(D) = -\int_{\infty}^{D} P(x)dx = 64 n_0 k_B T \gamma_0^2 (1/\kappa) \exp(-\kappa D).$$

$P(D)$ and $U_{el}(D)$ can be directly derived from experimental measurements using the Israelashvili surface force apparatus (Figure 2.23, Sections 2.3 and 2.5).

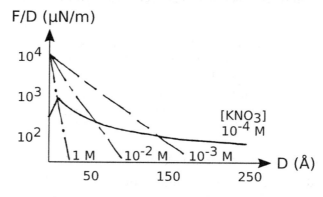

FIGURE 2.23 Measurement of the interaction force as a function of the distance separating two mica plates immersed in an electrolyte ($KNO_3$).

### 2.4.4 DLVO Theory: Stability of Colloidal Suspensions and Soap Films

#### 2.4.4.1 Colloidal Particles

Let us start by considering a suspension of, say, negatively charged colloidal particles. The interaction energy $U(D)$ is the sum of the repulsive electrostatic energy $U_{el}$ (see above) and the attractive van der Waals energy $U_{vdW}$ (Section 2.3). To set the ideas down, we suppose that the colloidal particles have a radius $R \gg D$.

$$U(D) = -\frac{A}{6}\frac{R}{D} + \frac{64 n_0 k_B T \gamma_0^2}{\kappa} e^{-\kappa D}$$

with $A$ the Hamaker constant (Section 2.3).

While the attractive van der Waals term only depends on the distance $D$, the repulsive contribution to the energy varies as $\kappa^{-1} \exp(-\kappa D)$, and is thus dominated by the exponential variation $\exp(-\kappa D)$.

The graph in Figure 2.24 qualitatively represents the variation of $U$ with distance for different salt concentrations, hence different values of the screening length $\kappa^{-1}$.

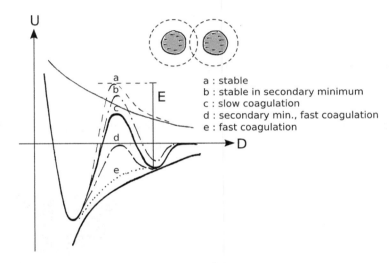

FIGURE 2.24   DLVO theory applied to colloidal stability (or coagulation).

- At weak ionic forces, electrostatic repulsion dominates. The potential $U(D)$ has a primary minimum and a secondary minimum separated by an energy barrier $E \gg k_B T$. The colloidal suspension is stable, and the average distance between particles is set by the position of the secondary minimum.

- As the salinity of the medium increases, the range of the electrostatic repulsion decreases, as well as the activation barrier $E$. When $E \sim k_B T$, the colloidal particles may escape the secondary well under thermal agitation and are attracted to the primary potential well controlled by van der Waals interactions. This is the coagulation process.

- At very high ionic strength, electrostatic screening occurs only over atomic distances and the interaction potential is dominated by the attraction of van der Waals forces. The coagulation is then very fast.

Note that in the latter case, we obtain "fractal" aggregates (Section 2.1 and Figure 2.25). The number of colloids in a sphere of radius $r$ is: $n(r) = r^{Df}$, where $D_f$ is the fractal dimension (Section 2.1) [2].

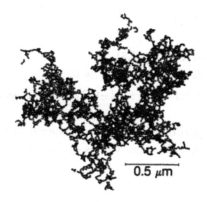

FIGURE 2.25 Electron transmission micrograph of a colloidal aggregate obtained by DLA (Diffusion Limited Aggregation) (4,739 particles). $D_f = 1.77$. (Extracted from [3].)

### 2.4.4.2 Soap Films

The same DLVO approach can be applied to discuss the stability of soap films, which consist of two monolayers of surfactants separated by a film of water of thickness $e$ (Figure 2.26).

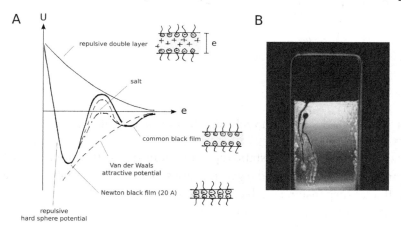

FIGURE 2.26 **(See color insert.)** (A) Classification of soap films from DLVO theory. (B) Draining of a vertical film that becomes a black film. (Photo by K. Mysels.)

- Thick soap films ($e > 100$ nm) can be obtained at low electrolyte concentrations, for which electrostatic repulsions exceed the van der Waals attraction. These films exhibit iridescence due to interferences between white light rays reflected on each interface.

At intermediate salinities, a "common black film" is stable at thicknesses of the order of 10 to 100 nm, corresponding to the secondary minimum.

- If salinity increases or if there is drainage, the van der Waals forces become dominant. A "Newtonian black film" with a thickness of the order of 2 nm is obtained, stabilized by short-range hard-core interactions (which are not explicitly present in the DLVO theory).

References

1. J. Israëlachvili, *Intermolecular and Surface Forces*. Academic Press, 1985.
2. T.A. Witten, L.M. Sander, *Phys. Rev. Lett.*, 47, 1400 (1981).
3. D.A. Weitz, M. Oliveria, *Phys. Rev. Lett.*, 52, 1433–1436 (1984).

## 2.5 MICROMANIPULATION AND MICROFLUIDICS

Numerous techniques have been developed to analyze and manipulate soft matter objects at scales ranging from nanometers to microns. Lab-on-Chips were created using approaches taken from microelectronics. This led to the creation of a new discipline, namely microfluidics. We give here a brief and non-exhaustive description of some experimental micromanipulation methods adapted to the microscopic and mesoscopic world.

### 2.5.1 Force Probes

#### 2.5.1.1 Surface Force Apparatus (SFA) [1, 2]

This setup measures the interaction between two surfaces, which are smooth at the atomic scale and decorated with molecules (polymers, surfactants, colloids). The separation distance, probed by an interferometric technique, can be controlled and measured with an accuracy of a few Å to μm. This approach allowed the characterization of van der Waals interactions and electrostatic or steric repulsions. Israëlachvili apparatus (Figure 2.27) is delicate to operate, and its use mostly remained confined to academic groups.

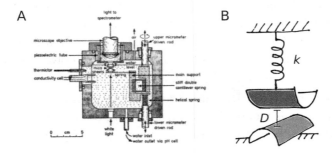

FIGURE 2.27 (A) Drawing of the Surface Force Apparatus. (Adapted from [2].) (B) Sketch of interacting half-cylinders.

### 2.5.1.2 Atomic Force Microscope (AFM) [3]

Atomic force microscopy is widely used in both academia and industry. This technique allows the measurement of the normal and tangential forces between the tip connected to a cantilever and the substrate (Figure 2.28A). AFM has an atomic resolution microscope in that tip scanning allows the reconstruction of the topography of the substrate, the visualization of adsorbed proteins, and tracking of their conformational changes under the action of light or drugs.

A "macro" version of the AFM (Figure 2.28B), where the tip is replaced by a millimetric elastic bead, has been developed to investigate the adhesion and friction of beads in the air or immersed in a liquid, as well as the dewetting of very thin liquid films, thus mimicking aquaplaning between a tire and the wet road. Force measurement derived from cantilever deflection was combined with reflection interference contrast microscopy (RICM) to monitor the evolution of the contact.

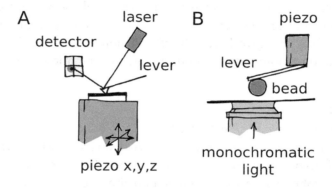

FIGURE 2.28 (A) Main components of an AFM. (B) Version of the macro-AFM. (A. Buguin, Institut Curie.)

### 2.5.1.3 Optical Tweezers [4]

A bead of refractive index greater than the index of refraction of water and with a diameter between 10 nm and 10 μm can be trapped at the focal point of a laser beam as a result of the balance between the force due to the intensity gradient laser and the radiation pressure (Figure 2.29A). This optical trap allows the manipulation of polymers, proteins, or microorganisms. For instance, Figure 2.29B shows a DNA chain grafted on the trapped bead. As the sample moves, the polymer deforms under the flow of uniform velocity $\vec{V}$. The optical trap allows the measurement of weak forces over a range of 1 to 100 picoNewtons. There are also magnetic traps, where magnetic beads are manipulated by a magnetic field gradient.

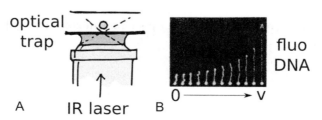

FIGURE 2.29 (A) Optical tweezers. (B) DNA elongation. (Courtesy of Steve Chu.)

### 2.5.1.4 Micropipettes and BFP (Biomembrane Force Probe) [5]

A bead decorated with adhesion proteins A is glued to a red blood cell that serves as a force sensor (Figure 2.30). Another bead is decorated with complementary proteins B which bind to A (key-lock). This technique allows the study of the interaction between two proteins and more importantly the measurement of the strength of the AB complex.

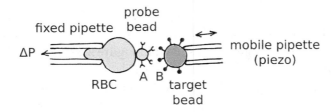

FIGURE 2.30 Micropipettes and *biomembrane force probe*. The mobile pipette is moved with a piezoelectric translator and holds the "test" ball coated with protein B. A red blood cell is aspirated in the second pipette (suction pressure $\Delta P$). A bead functionalized with protein A is glued to the red blood cell.

### 2.5.2 Microfluidics – MEMS and Lab-on-a-Chips

This field describes the miniaturization of electromechanical and fluidic systems at submicron scales, with two important advantages: decreasing volumes and analysis times. In the 1980s, MEMS (Micro Electro Mechanical Systems) could detect and process a signal on a chip, which can be duplicated in millions of copies. For instance, the sensor contained in the airbag of a car is a MEMS accelerometer. Since the 1990s, MEMS have been used in chemistry and biology and led to the emergence of a new discipline: "microfluidics," which generally deals with the flow of simple or complex liquids in artificial microsystems.

These systems allow the carrying out of multiple operations (Figure 2.31), such as detecting biological molecules or cells, displacing and sorting them from a raw sample, following a chemical reaction kinetics. Steve Quake's laboratory (Stanford University) has pioneered the design of these microfluidic systems [6]. The microfabrication of channels, pumps, and

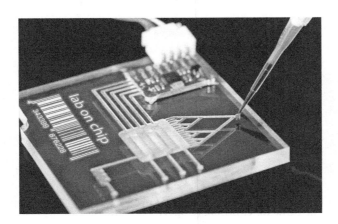

FIGURE 2.31 Micrograph of a Lab-on-a-chip.

valves is based on the use of elastomers (such as PDMS). Chips are barely larger than a coin. They can contain up to several hundred compartments and valves to actuate liquids.

Microfluidics may have more surprising applications, such as in children's games. We all have played these maze games to help a mouse find the right path to the cheese when we were kids. The one shown in Figure 2.32A is particularly simple. When it becomes more complicated, microfluidics can help us.

George Whitesides, a pioneer in soft lithography for the fabrication of microfluidic networks, and his collaborators have molded a labyrinth with PDMS (poly-dimethyl siloxane). The principle of soft lithography is shown in Figure 2.32B. We begin by depositing a film of liquid resin that is baked and solidified on a glass slide or silicon wafer (less rough). A silica mask on which are printed patterns (by metal deposition) is then deposited on the resin film. By irradiating the surface with UV light, the unprotected resin becomes soluble in the developing solvent if it is said to be "positive" (respectively insoluble if it is said to be "negative"). Therefore, a relief surface is formed in the photosensitive resin. Finally, by pouring liquid PDMS with a chemical crosslinking agent, a transparent and flexible elastomer is formed which conforms to the printed shapes. In the end, the PDMS mold obtained reproduces the negative of the initial 3D shape. To produce microfluidic mazes, patterns are lines [7]. In the case of Figure 2.32C, there is only one possible path between the input (top) and the output (bottom). For better visualization, the researchers first filled all the channels by plugging the outlet with a liquid that wets PDMS (Sections 4.3, 4.4, 4.5). If a channel has a dead end, the air, which is initially trapped, will eventually diffuse (because PDMS is permeable to gases), and the liquid fills the entire maze. Then, the injection of another liquid, immiscible (with the first one) and stained, moves the first fluid, except in dead ends because the liquids are incompressible and therefore select the path that leads to the exit.

More difficult, if there are several possible paths, the liquid will follow "the best one," or more precisely the one with "less resistance." There is indeed a direct analogy between microfluidics and electricity. The electric current $i$, which is associated with the movement of the electrons, corresponds to the (volumetric) flow rate of the fluid $Q$. In the same way as $i = \iint_S \vec{j} \cdot d\vec{S}$ (with $j$ the current density and $S$ the section of the conductor), $Q = \langle v \rangle \cdot S$,

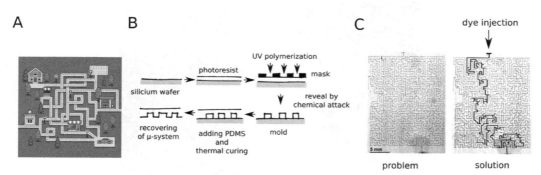

FIGURE 2.32 (A) Maze game for children. (Copyright Shutterstock.) (B) Principle of PDMS-based soft lithography. (C) Microfluidic maze. (Adapted from [7].)

where $\langle v \rangle$ is the average speed of the fluid (which has a Poiseuille profile: it is maximum in the center of the channel and vanishes on the walls). The analog of the voltage $U$ is the pressure drop or pressure difference $\Delta P$ between the input and the output. Finally, we know the Ohm law that defines the electrical resistance $R_E$: $U = R_E.i$. The microfluidic analog is $\Delta P = R_H.Q$, with $R_H$ the hydraulic resistance.

According to Hagen-Poiseuille's law, for a channel of length L and of circular cross-section of radius $R$ and for a liquid of viscosity $\eta$, we have: $\Delta P = \left(8L\eta / \pi R^4\right)Q$. This expression shows that, for a fixed pressure difference, the smaller the length or the larger the radius of the channel, the higher the flow (or the shorter the time spent in the channel). To return to our maze, one might think that it is relatively easy to guess the best path by determining the one of shorter length. However, as with electrical resistors, there may be parallel "shunts" or resistors that affect the flow on the main path. On the other hand, the role of the channel section (height × width for rectangular section channels) is less easy to grasp. Taking into account the number of lanes in Boston's network of roads and highways through the width of the channels, this fluidic method was used to find the fastest paths, without having to use a GPS.

One of the limitations of microfluidics is in the fabrication of three-dimensional microfluidic structures, which requires precise alignments of the different layers. In recent years, with the explosion of the use of 3D printers, molds are now manufactured in 3D. Very complex shapes can be designed and allow the association of several modules fulfilling different microfluidic functions. In order to allow large combinations of modules, the group of A. P. Lee at UC Irvine [8] has taken inspiration from Lego® games to create microfluidic platforms. By printing molds that are the negatives of Lego blocks, they have made PDMS structures, which can be assembled horizontally and vertically (while ensuring a good seal thanks to PDMS elasticity and hydrophobicity), to assemble complex fluidic circuits.

## References

1. J. Israëlachvili, D. Tabor, *Proc. R. Soc. Lond. A*, 331, 19–38 (1972).
2. J. Wong, A. Chilkoti, V.T. Moy, *Biomol. Eng.*, 16, 45–55 (1999).
3. G. Binnig, C.F. Quate, Ch. Gerber, *Phys. Rev. Lett.*, 56, 930 (1986).
4. Ashkin et al. *Opt. Lett.*, 11, 288–290 (1986).
5. E. Evans et al. *Biophys. J.*, 68, 2580–2587 (1995).
6. J.W. Hong, S.R. Quake, *Nat. Biotechnol.*, 21, 1179–1183 (2003).
7. M.J. Fuerstman et al. *Langmuir*, 19, 4714–4722 (2003).
8. K. Vittayarukskul, A.P. Lee, *J. Micromech. Microeng.*, 27, 035004 (2017).

# CHAPTER 3

# Phase Transitions

## 3.1 PHYSICAL TRANSFORMATIONS OF PURE SUBSTANCES

An ice cube melts and water evaporates when heated. The solid becomes a liquid, which then turns into vapor, and all these transformations are reversible. By definition, the *phase* of a substance refers to a chemically and physically homogeneous state of matter. A substance may have solid, liquid, and gas phases and often polymorphic solid phases such as graphite and diamond for carbon. A *phase transition* is the spontaneous conversion of a phase into another. For a given pressure it occurs at a characteristic temperature. At atmospheric pressure, ice melting occurs at 0°C and water boiling at 100°C. The transition temperature is defined by the equality of the chemical potential of a given element in the two phases. A *phase diagram* (Figure 3.1) is a chart that defines the conditions or physical parameters, such as temperature and pressure, at which the different phases are present or coexist.

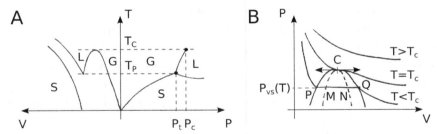

FIGURE 3.1    (A) Phase diagram in $(T,P)$ coordinates and in $(T,V)$ coordinates; (B) Compression isotherm of a gas in coordinates $(P,V)$. Notations: $T$ = temperature, $P$ = pressure, $V$ = volume, $S$ = solid phase, $L$ = liquid phase, $G$ = gas phase.

All phase diagrams can be discussed from a general principle derived by J.W. Gibbs, [1] namely the phase rule:

$$v = C + 2 - \varphi$$

where $v$ is the variance, i.e. the number of intensive variables (pressure, temperature, …) that can be changed independently without changing the number of phases, $C$ is the number of components, i.e. the number of chemical species, and $\varphi$ is the number of phases.

### 3.1.1 One-Component Systems

We will begin with the simple case of one component ($C=1$), which exists only in three phases according to the phase rule:

*Crystal*: a molecular crystal is a condensed solid state, characterized by the long-range order of the position of the molecules that oscillate around an equilibrium position. The structure is invariant by discrete rotations and translations.

*Liquid*: a liquid is also a condensed state but with a disorder of the center of mass of the molecules. It is invariant by continuous rotations and translations.

*Gas*: a gas is a non-compact fluid state, with the same symmetries as a liquid.

#### 3.1.1.1 Phase Diagrams (P,V,T)

Discussion of the phase diagram shown in Figure 3.1:

At the coexistence between two phases ($\varphi = 2$), the variance is $v=1$ according to the phase rule.

**In coordinates $P,T$** (Figure 3.1A left), $v=1$ means that the pressure $P$ is a function of the temperature $T$: the curves $P(T)$ are the coexistence curves. On the coexistence curves, all the heat that is received (or transferred) is used to perform the phase change at constant $T$. The slope of the solid–liquid coexistence curve (fusion) is generally positive, except for $H_2O$, because ice is less dense than water. The slope of the liquid–vapor coexistence curve (boiling) is generally positive, as illustrated by Franklin's broth experience. The solid–vapor coexistence curve corresponds to sublimation.

**In coordinates P,V,** we show in Figure 3.1B isotherm compression $P(V)$ at fixed temperature $T$ of a compound initially in a gaseous state. When $T<T_c$, a droplet of liquid is nucleated at $V=V_Q$. For $V_P<V<V_Q$, the gas is gradually transformed into a liquid: it liquefies. Note that at the G/L coexistence, the saturated vapor pressure $P=P_{VS}(T)$ is fixed. For $V=V_P$, all the gas is liquefied, and for $V<V_P$, the compound is in its liquid phase. For $T=T_c$, the plateau $P$–$Q$ (Figure 3.1B) reduces to a point C, the critical point, where liquid and gas phases become identical and this state is called supercritical fluid. The system becomes opalescent because of large density fluctuations at the critical point. For $T \gg T_c$, the gas becomes a perfect gas ($PV \cong RT$ per mole).

Two points in the phase diagram are remarkable:

*Triple point*: three phases coexist and $v=0$. The three phase boundaries meet at the triple point. For water, $T_t=273.16=K=0.01°C$; $P_t=6 \times 10^{-3}$ atm.

*Critical point*: there is a continuous phase transition between a homogeneous liquid phase and an L/G biphasic phase. Above the critical temperature $T_c$, a single uniform phase called supercritical fluid fills the container and interfaces between phases disappear. For water, $T_c=374°C$; $P_c=218$ atm.

### 3.1.1.2 Thermodynamics of Phase Separation: Free Enthalpy and Chemical Potential

For a system at $P$ and $T$ constant, which is the case for many transformations at atmospheric pressure, the free enthalpy $G = U - TS + PV$ is minimal.

Let us consider a pure substance present in two phases 1 and 2 in a container maintained at $T$ and $P$ constant. As the free enthalpy is an extensive function, $G = G_1 + G_2$. We note $n_1$, $n_2$ the number of molecules in phases 1 and 2 respectively, with $n = n_1 + n_2$. $G$ is then written:

$$G(P,T,n_1) = n_1\mu_1 + n_2\mu_2 = n_1(\mu_1 - \mu_2) + n\mu_2 \tag{3.1}$$

where $\mu_i$ are the chemical potentials in the phases $i = 1$, 2. They are defined as the free enthalpy per molecule. Thermodynamic equilibrium is given by:

$$dG/dn_1 = 0, \text{ i.e. } \mu_1(P,T) = \mu_2(P,T)$$

This equation leads to the following rule: *at equilibrium, the chemical potentials of a substance are equal in all phases.*

Using the thermodynamic relationship $d\mu_i = v_i dP - s_i dT$, where $v_i$ is the molecular volume and $s_i$ the molecular entropy, the condition $d\mu_1 = d\mu_2$ along the coexistence curve leads to:

$$dP / dT = (s_2 - s_1) / (v_2 - v_1). \tag{3.2}$$

The *latent heat* $L_{1\to2}$ is the heat exchanged at constant temperature during the phase transition to change a mass $m$ from phase 1 to phase 2. From the second law of thermodynamics, $s_2 - s_1 = L_{1\to2}/T$. $L$ is expressed in J kg$^{-1}$ or J mol$^{-1}$. For water, the latent heat of fusion and evaporation are respectively $L_f = 334$ kJ kg$^{-1}$ (or 334 J to melt 1 g of ice) and $L_v = 2{,}257$ kJ kg$^{-1}$ (or 2,257 J to evaporate 1 g of water). Evaporation of water on our skin is used to cool our body in summer.

Equation 3.2 leads to the *Clapeyron formula*:

$$dP / dT = L_{1\to2} / T(v_2 - v_1) \tag{3.3}$$

At the liquid/vapor coexistence, $s_{vap} > s_{liq}$ and $v_{vap} > v_{liq}$, which leads to $dP/dT > 0$.

As a consequence, *the slope $dP/dT$ of the L/V coexistence curve is always positive.* This is illustrated by the Benjamin Franklin (1706–1790) experiment described below.

*The slope of the S/V coexistence curve is always positive* because $v_{vap} > v_{sol}$.

*The slope of the S/L coexistence curve is always positive, except for water.* This is illustrated by the John Tyndall (1820–1893) experiment shown in Figure 3.2B and described hereafter.

For the S/L coexistence curve, in most cases, $v_{liq} > v_{sol}$, and the slope is positive. But there is a significant exception for water, for which $v_{sol} > v_{liq}$. This explains the explosion of bottles of water in the freezer and the fact that ice cubes float. This leads to $dP/dT < 0$. An ice cube at a temperature below 0°C melts under compression ($P_f = 3$ atm at $T = -5$°C).

### 3.1.1.3 Benjamin Franklin Boiling Water: How to Boil Water with Cold Water [2]

This experiment can be performed in our kitchen with a glass container (jam jar) and a sponge as shown in Figure 3.2A.

Half fill the container with water; heat to boiling and keep it boiling for three minutes. Remove from the hot plate, close the jar with the lid, and flip it over: the water stops boiling. Then wipe the top of the container with the wet sponge: boiling resumes. Remove the wet sponge: boiling stops. As soon as you cool down again, the water starts to boil again. When water boils, liquid water turns into water vapor, which expels the air initially present in the container and then occupies half of the container. When you remove from the heat, the boiling stops so the water temperature is below 100°C. With the sponge the water vapor is cooled, which condenses and becomes liquid again; as the container is plugged, the upper half of the container is almost empty, which means that the pressure above the water is then very low.

It shows that the lower the pressure above the water, the lower the boiling temperature of the water; when the boiling temperature of water has been sufficiently lowered, it will boil again but will stop very quickly if you do not continue to condense the water vapor that forms.

### 3.1.1.4 Tyndall Experiment on the Fusion of Ice [3]

A wire carrying weights at both ends is placed on a block of ice as shown in Figure 3.2B. Under the action of pressure, the melting temperature drops locally and the ice melts, causing the wire to sink slowly into the ice. Released from the pressure, the water formed immediately freezes. The wire thus passes through the entire ice block without cutting it. John Tyndall was an Irish physicist interested in glaciological problems. His studies on ice melting and freezing allowed him to explain the movement of glaciers (1871).

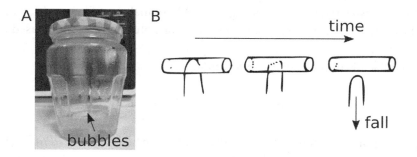

FIGURE 3.2 (A) Franklin boiling water experiment in the kitchen. (Photograph by FBW.) (B) Drawing of Tyndall experiment on the fusion of ice.

## 3.1.2 Binary Mixtures

When two components $A$ and $B$ ($C = 2$) are present in the system, $v = 4 - \Phi$. If the pressure is held constant, the phase diagram is depicted in terms of temperature and composition. Let us consider the phase diagram of water ($A$) and phenol ($B$) at atmospheric pressure. $\Phi$ is the volume fraction of $A$ (water) defined as $\Phi = ca^3$, where $c$ is the concentration of $A$ and $a^3$ the molecular volume (Figure 3.3).

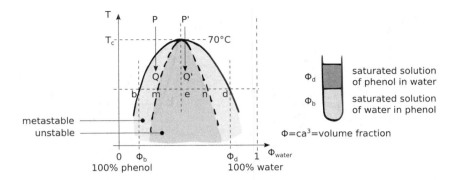

FIGURE 3.3 Phase diagram of phenol/water mixtures. The coexistence curve separates monophasic and biphasic phases (dark grey region). The spinodal curve separates metastable and unstable domains. Point b: limit of solubility of water in phenol. Point d: limit of solubility of phenol in water. Point e: separation in two phases of composition $\Phi_b$ and $\Phi_d$. For $T > T_c$, the phenol is miscible in all proportions. When the system is cooled from (1) P to Q, one observes a phase separation by nucleation and growth of droplets, (2) from $P'$ to $Q'$, the fluctuations of concentration are amplified (spinodal decomposition).

Near the critical point $(T_c, \Phi_c)$, the large concentration fluctuations give rise to intense light scattering named critical opalescence. The interface between the two phases observed below $T_c$, rich in water and phenol respectively, disappear progressively. This is a case of a second order phase transition as discussed in Section 3.2.

### 3.1.3 Analogies between Liquid–Gas Transition and A/B Phase Separation

As seen in Figure 3.4, there is a complete analogy between the van der Waals isotherms $P(V)$ of a gas compression (Figure 3.4A) and the exchange chemical potential μ versus composition for an *A/B* binary mixture (Figure 3.4B). In both cases, the plateau of co-existence between a liquid and a gas phase, or a dilute and a concentrated solution, is given by the equality of the area of hatched zones to satisfy the equality of the chemical potential $\mu_I = \mu_{II}$ and the pressure in the two phases.

The criterium of thermodynamic stability for a gas is $dP/dV < 0$ and for the *A/B* mixture $d\mu/dc > 0$: the branch MN is unstable and the system separates in two phases. The gas phase is metastable on the path NQ (supersaturated vapor) and the liquid phase on the path PM (superheated liquid).

The isotherms shown in Figure 3.4 were first described by van der Waals for the one component case and by Flory–Huggins for the binary mixture. They are both mean-field theories, ignoring the large fluctuations of the system near the critical point.

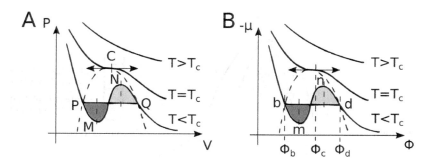

FIGURE 3.4 (A) Van der Waals isotherms of gas compression: gas liquefaction appears below $T_c$. (B) Exchange chemical potential $\mu = \mu_a - \mu_b$ versus composition for a binary mixture. Above $T_c$, the solution is homogeneous. Below $T_c$, the solution separates in two phases. The interfacial energy between the two phases vanishes at $T_c$.

### 3.1.3.1 Van der Waals Model of L/G Transitions [4]

The state equation of an ideal gas is $PV = RT$ (per mole), where $R$ is the gas constant ($R = 8.31441$ J K$^{-1}$ mol$^{-1}$). The ideal gas law makes the assumption that gas particles have no volume and are not attracted to each other. By including the interaction potential between molecules, which is attractive at long distances and repulsive at short distances (hard sphere potential), one has to integrate two modifications:

$V$ reduces to $V - b$, where $b$ is the molecular volume, and the pressure due to van der Waals interactions becomes $P + A/V^2$. Altogether, the state equation becomes:

$$(V - b)(P + A/V^2) = RT$$

The isotherms (Figure 3.4A) show that the system is unstable when $P' = (\partial P / \partial V)\big|_T$ is positive and separates in two phases. The critical point is given by $P' = P'' = 0$, i.e.:

$$\left(\partial P / \partial V\right)\big|_T = \left(\partial^2 P / \partial V^2\right)\big|_T = 0$$

$$\frac{\partial P}{\partial V}\bigg|_T = \frac{-RT_c}{\left(V_c - b\right)^2} + \frac{2A}{V_c^3} = 0$$

$$\frac{\partial^2 P}{\partial V^2}\bigg|_T = \frac{2RT_c}{\left(V_c - b\right)^3} - \frac{6A}{V_c^4} = 0$$

It leads to:

$$T_c = \frac{8A}{27BR}$$

$$V_c = 3B$$

$$P_c = \frac{A}{27B^2}$$

For water: $T_c = 647$ K, $V_c = 56$ cm$^3$, and $P_c = 218$ atm.

### 3.1.3.2 Flory–Huggins Model of A/B Mixtures [5, 6]

When $N_A$ molecules of $A$ are mixed with $N_B$ molecules of $B$, the free energy of mixing $G$ has an entropic and an enthalpic contribution (Figure 3.5). The two terms can be calculated in a lattice model: each lattice site is either occupied by one molecule $A$ or one molecule $B$, as shown in Figure 3.5. We denote $\Phi$ the fraction of sites occupied by molecule $A$, which is related to the concentration through $\Phi = ca^3$, where $a^3$ is the volume of a unit cell in the cubic lattice. The entropy term describes how many arrangements of A exist on the lattice for a given $\Phi = N_A/(N_A+N_B)$, and the energy term describes the interactions between adjacent molecules. It leads to:

$$G(\Phi) = (N_A + N_B)k_B T\left[\Phi Ln\Phi + (1-\Phi)Ln(1-\Phi) + \chi\Phi(1-\Phi)\right]$$

$$= (N_A + N_B)G_{site}$$

with $\chi k_B T = z\Delta h$, $z$ = number of neighbors, $G_{site}$ the free enthalpy per site, and $\Delta h$ defined in Figure 3.5.

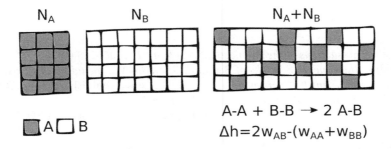

$$N_A \qquad N_B \qquad N_A+N_B$$

$$\square A \square B$$

$$\text{A-A + B-B} \rightarrow \text{2 A-B}$$

$$\Delta h = 2w_{AB} - (w_{AA} + w_{BB})$$

FIGURE 3.5   Lattice model for AB mixtures used to calculate the entropy and the enthalpy of mixing. $\Delta h$ is the enthalpy per new A–B bond formed by breaking A–A and B–B bonds.

From $G(\Phi)$, using $N_A$ and $N_B$ as variables, one can calculate the chemical potentials $\mu_A$, $\mu_B$ of $A$ and $B$ defined by:

$$dG = \mu_A dN_A + \mu_B dN_B$$

with $\mu_A = \left(\partial G/\partial N_A\right)\big|_{N_B,T,P}$ and $\mu_B = \left(\partial G/\partial N_B\right)\big|_{N_A,T,P}$
or, using $N_a$ and the volume $V = (N_a + N_b)a^3$ as variables, noting $\mu = \mu_a - \mu_b$ the exchange chemical potential, and $\Pi$ the osmotic pressure:

$$dG = \mu dN - \Pi dV$$

with $\mu = \left(\partial G/\partial N_A\right)\big|_{V,T,P} = \dfrac{\partial G_{site}}{\partial \Phi} = \mu_A - \mu_B$ and $\Pi = -\left(\partial G/\partial V\right)\big|_{N_A,T,P} = \dfrac{1}{a^3}\Phi^2\dfrac{\partial\left(G_{site}/\Phi\right)}{\partial\Phi}$

From the free energy of mixing $G$, one can derive the phase diagram. The stability criterium is $G'' > 0$, and $G'' = 0$ defines the spinodal.

$\mu' = \mu'' = 0$ defines the critical point as shown in Figure 3.4B:

$$\frac{\mu'}{k_B T} = \frac{1}{\Phi} + \frac{1}{1-\Phi} - 2\chi = 0$$

$$\frac{\mu''}{k_B T} = -\frac{1}{\Phi^2} + \frac{1}{(1-\Phi)^2} = 0$$

The critical point is thus characterized by: $\Phi_c = 1/2$ and $\chi_c = 2$. The spinodal is given by:

$$1/\Phi + 1/(1-\Phi) = 2\chi$$

The coexistence curve is deduced from the equilibrium relationship:

$$\left. \begin{array}{l} \mu_A(I) = \mu_A(II) \\ \mu_B(I) = \mu_B(II) \end{array} \right\} \quad \text{or} \quad \left. \begin{array}{l} \mu(I) = \mu(II) \\ \Pi(I) = \Pi(II) \end{array} \right\}$$

### References

1. W.J. Gibbs, *Transactions of the Connecticut Academy*, III, 108–248 (1876).
2. B. Franklin, *The Complete Works of Benjamin Franklin V5: Including His Private as Well as His Official and Scientific Correspondence (1887)*. Kessinger Publishing, LLC (2010).
3. J. Tyndall, T.H. Huxley, *Proc. R. Soc. Lond.*, 8, 331–338 (1857).
4. J.D. van der Waals, On the Continuity of the Liquid and Gaseous States, PhD thesis, University of Leiden (1873).
5. M.L. Huggins, *J. Phys. Chem.*, 46, 151 (1942).
6. P.J. Flory, *Principles of Polymer Chemistry*. Ithaca, NY: Cornell University Press, 1953.

## 3.2 CRITICAL PHENOMENA: FROM FERROMAGNETISM TO LIQUID CRYSTALS

Since the discovery of the critical point of the carbon dioxide liquid–gas transition, similar phenomena have been observed in many systems such as magnetic materials, superconductors, superfluid helium, and liquid crystals. The work on magnetic transitions led to a universal description of phase transitions and contributed to the glorious era of solid-state physics. Each transition is characterized by an order parameter of dimension $n$. References 1 and 2 are excellent introductions to critical phenomena and phase transitions in liquid crystals.

### 3.2.1 Magnetic Transitions: Order Parameter and Critical Exponents

Let's take an iron rod and heat it up to red: the iron is not magnetized; it is said to be in a paramagnetic state. At the microscopic level, the magnetic moments $\vec{\mu}$ of the different iron atoms are randomly oriented (Figure 3.6A). If a magnetic field $\vec{H}$ is applied, a magnetization vector $\vec{M}$ (defined per unit volume) $\vec{M} = \chi_T \vec{H}$ is induced, where $\chi_t$ is the magnetic susceptibility first calculated by Curie ($\chi_t \propto T^{-1}$) (Figure 3.6A).

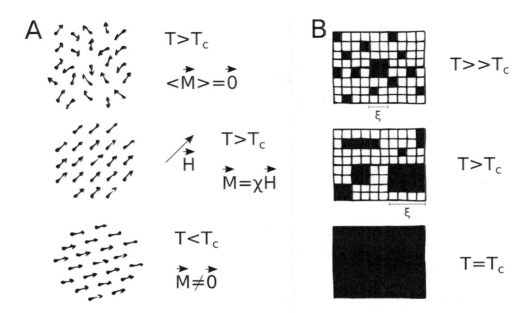

FIGURE 3.6 Magnetic transitions. (A) Paramagnetic-ferromagnetic transition: $\overline{M}$ is a verctor ($n = 3$). Above $T_c$, $\overline{M} = 0$ and $\overline{M}$ can be induced by a magnetic field H. Below $T_c$, the magnetic moments align spontaneously. (B) Ising model ($n = 1$). The black(white) square is the spin up (down), above, near, and below $T_c$. The Ising model is a simplified mathematical description of phase transitions. The model consists of a lattice of spin, each of which interacts with its nearest neighbors. The model undergoes a transition from disorder (where the spins are more or less randomly aligned) to order (where the spins are aligned) at a certain temperature, called the critical temperature.

Now let's cool the sample. Below a certain critical temperature $T_c$, the critical temperature of the material, the system becomes magnetized. Microscopically, the elementary magnetic moments are aligned in a given direction (Figure 3.6A). There is now a long-range magnetic order. This state is characterized by an *order parameter*. For this ferromagnetic transition, we choose the magnetization vector $\overline{M}$. Since $\overline{M}$ is a vector defined by its three components, the dimension of the order parameter is $n = 3$ and the space dimension is $d = 3$. It corresponds to the Heisenberg model [1].

The variation of $M$ with temperature is shown in Figure 3.7A. $M(T)$ vanishes at $T_c$, and the transition is referred to as *second order transition*. Close to the critical point, the system can switch to several states: here, a state of null magnetization and those with a "finite" magnetization. As a result, the system is very sensitive to external disturbances. The field coupled to the magnetization is the magnetic field. When the temperature is lowered, as the critical temperature is approached, the susceptibility becomes very high and diverges at $T_c$. At the microscopic level, magnetic moments are already parallel to each other within domains. The size of these domains is called the correlation length $\xi$ and can be determined by neutron scattering measurements. The correlation length $\xi$ diverges at $T_c$ (Figure 3.8).

To summarize, the ferromagnetic transition is characterized in the disordered state, above $T_c$, by a magnetic susceptibility and a correlation length that both diverge at $T_c$. Below $T_c$, in the ordered state, it is characterized by the measurement of the order parameter and by the way it vanishes at $T_c$.

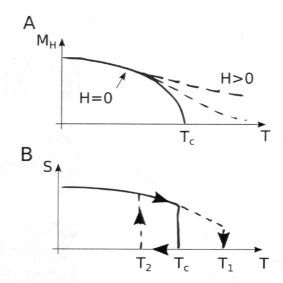

FIGURE 3.7 (A) Magnetization versus temperature at fixed $H$. $M_{H=0}$ vanishes at $T_c$, indicating a second order transition; (B) Nematic order parameter $S(T)$. $S(T)$ is finite at $T_c$, indicating a first-order transition. Hysteresis is characteristic of a first-order transition.

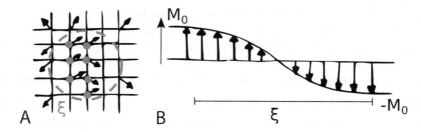

FIGURE 3.8 Correlation length: (A) Size of the order parameter fluctuations above $T_c$; (B) Size of the walls between ($\uparrow$) and ($\downarrow$) domains below $T_c$.

Magnetization in conventional magnetic materials can point in all three directions of space, and $n = 3$. Heisenberg's name is associated with this case, as he was the first to propose a quantum theory of coupling between elementary magnetic moments. However, in some materials the magnetization $\vec{M}$ is confined to a plane: it has only two components, and $n = 2$. This case is referred to the XY model, in the theoretical literature, and this model allows the description of other transitions for which $n = 2$. Finally, in some material systems magnetization is confined to a single direction (spin $+1/2, -1/2$), and $n = 1$ (Figure 3.6B). These systems are called Ising models and are of great importance because they also describe the L/G and L/L transitions. In the lattice models shown Figure 3.5 for binary mixtures, each site had two states: for the spins (up and down), for the L/G transition (occupied, empty), for binary mixtures (molecules A and B).

## 3.2.2 Definition of Critical Exponents
The critical exponent characterizes a power law $f(x)$. By definition:

$$\lambda = \lim_{\varepsilon \to 0} \left( \ln f(x) / \ln \varepsilon \right), \quad \varepsilon = (T - T_c) / T_c$$

We give in Table 3.1 the definition of critical exponents for magnetic and fluid systems.

TABLE 3.1 Definition of Critical Exponents for Magnetic and Fluid Systems

| Exponent | Magnetic System | Fluid System |
|---|---|---|
| $\beta$ | $M \sim (-\varepsilon)^{\beta}$ | $\rho_L - \rho_G \sim (-\varepsilon)^{\beta}$ |
| $\gamma$ | $\chi_T \sim (-\varepsilon)^{-\gamma}$ | $K_T \sim (-\varepsilon)^{-\gamma}$ |
| | $\chi_T \sim \varepsilon^{-\gamma}$ | $K_T \sim \varepsilon^{-\gamma}$ |
| $\nu$ | $\xi \sim (-\varepsilon)^{-\nu}$ | $\xi \sim (-\varepsilon)^{-\nu}$ |
| | $\xi \sim \varepsilon^{-\nu}$ | $\xi \sim \varepsilon^{-\nu}$ |

The variations of $M$, $\xi$, and $\chi$ can be expressed with power laws as a function of $\varepsilon$, the deviation from the critical temperature.

The exponent, called the *critical exponent*, is "$\beta$" for the order parameter, $M \sim (-\varepsilon)^{\beta}$ that vanishes at $T_c$, "$\nu$" for the correlation length, $\xi \sim \varepsilon^{-\nu}$, and "$\gamma$" for the isothermal susceptibility $\chi_T \sim \varepsilon^{-\gamma}$ defined from the free energy $G = U - TS - MH$.

$$M = -\left( \partial G / \partial H \right)\big|_T, \quad \chi_T = \left( \partial M / \partial H \right)\big|_T = -\left( \partial^2 G / \partial H^2 \right)\big|_T$$

### 3.2.3 Large Fluctuations Give Large Response: Fluctuation Dissipation Theorem

The large response to small perturbations is a signature of critical phenomena and soft matter, both characterized by large fluctuations. The linear response theory demonstrates that the linear response of a system to a small external perturbation is expressed in terms of the fluctuations of the system in thermal equilibrium. It leads to:

$$\chi_T = \frac{1}{k_B T} \left[ \langle M^2 \rangle - \langle M \rangle^2 \right]$$

We can check that this relation leads to the Curie formula for magnetic moments at high temperature. If $N$ is the number of magnetic moments $\mu$:

$$\langle M \rangle = \sum_i \langle \mu_i \rangle = 0 \quad \langle M^2 \rangle = \sum_{i,j} \langle \mu_i \mu_j \rangle = N\mu^2$$

$$\chi_T = \frac{N\mu^2}{k_B T}$$

### 3.2.4 Extension to Other Transitions

#### 3.2.4.1 *Liquid/Gas Transition*

For the L/G transition described in Section 2.5.1, the order parameter is the difference in density between the liquid and gas phases ($\rho_l - \rho_g$). The order parameter has $n = 1$ component.

$$\rho_L - \rho_G \sim \left(-\varepsilon\right)^{\beta}$$

The variable coupled to the density is the pressure $P$. The isothermal compressibility $K_T$ is defined from the free enthalpy $G = U - TS + PV$. Note that the correspondence between fluid and magnetic systems is $V \rightarrow -M$ and $P \rightarrow H$.

$$K_T = -\frac{1}{V}\frac{\partial V}{\partial P}\bigg|_T = -\frac{1}{V}\frac{\partial^2 G}{\partial P^2}\bigg|_T$$

The *density–density* correlation function $g(r)$ is the correlation of the fluctuation of the densities from their average values at $\vec{0}$ and $\vec{r}$ and can be defined as:

$$g(r) = \left\langle \rho(\vec{0})\rho(\vec{r}) - \rho^2 \right\rangle \approx \frac{1}{a^2 r}\exp\left(-\frac{r}{\xi}\right)$$

The Fourier transform of $g(r)$ is measured by light or neutron scattering, leading to the measurement of $\xi(T)$.

#### 3.2.4.2 *Liquid Crystals Transitions*

Some molecules in the shape of rigid rods show intermediate order between a liquid and a crystalline state, as shown in Figure 3.9.

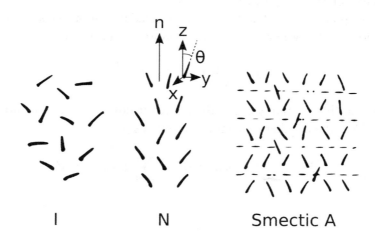

FIGURE 3.9  Liquid crystal transitions: isotropic–nematic and nematic–smectic A.

3.2.4.2.1 Nematic-isotropic transition   In nematics, the molecules are aligned along a preferred direction $\vec{n}$ (Figure 3.9 The states $n$ and $-n$ are identical and the order parameter $S$ is a tensor [2]:

$$S_{ij} = \langle n_i n_j \rangle - \frac{1}{3}\delta_{ij}$$

In the reference frame of the preferred axis $\bar{n}$ parallel to the z axis, we thus have:

$$\bar{S} = S(T)\begin{bmatrix} -\dfrac{1}{3} & 0 & 0 \\ 0 & -\dfrac{1}{3} & 0 \\ 0 & 0 & \dfrac{2}{3} \end{bmatrix} \quad \text{with } S(T) = (1/2)\langle 3\cos^2\theta - 1\rangle$$

$S(T)$ measures the alignment along $\bar{n}$ and is named the scalar order parameter. We show the curve $S(T)$ in Figure 3.7. At the nematic-isotropic transition, $S(T)$ does not vanish and is discontinuous. The transition is *first order* and not characterized by critical exponents. However, it is a weak first-order transition, and relatively large fluctuations of orientation are observed near $T_c$. These groups of molecules fluctuating along a preferred direction have been named *swarms* [3].

3.2.4.2.2 Smectic A/Nematic Transition   The high temperature nematic phase has a uniform density. The low temperature phase smectic A has a layered structure, meaning that the density is modulated in space. To characterize the smectic order it is necessary to know two parameters: the amplitude and phase of this modulation. The smectic/nematic order parameter has two components ($n = 2$), and it vanishes at $T_c$ (Figure 3.9).

Giant fluctuations of this order parameter are observed near $T_c$, a signature of second order transitions. Theoretically this transition has been described by Pierre-Gilles de Gennes (1972) thanks to the remarkable analogy he has drawn between superconductors and smectics A [4].

### 3.2.4.3 Superfluid Helium

The superfluid transition of helium-4 is a quantum phenomenon. The liquids on the two sides of this phase transition are usually referred to as He I and He II. At high temperatures (He I), it is a normal liquid. At low temperatures (He II), it flows without friction: it is a superfluid. The transition between He I and He II is a critical point, named a lambda point because the curve representing the specific heat of helium as a function of temperature resembles the Greek letter $\lambda$. The origin of this transition is due to the fact that He-4 is a boson, with a nucleus that contains two protons and two neutrons. In Bose statistics, each state can be occupied by any number of particles. At high temperatures, each state is occupied by at most a few particles, and He I is a regular liquid. At low temperatures, all particles occupy the same state of lower energy. This state can be described by a wave function, defined by an amplitude and a phase, which vanishes at $T_c$. In this case, the order parameter has $n = 2$ components in analogy with the magnetic transition, when $\vec{M}$ has two components.

### 3.2.5 Conclusion

All phase transitions are characterized by an order parameter.

If this parameter is discontinuous at $T_c$, the transition is *first order*.

If the order parameter vanishes at $T_c$, the transition is *second order*, where critical phenomena and giant fluctuations are observed. The critical exponents are a function of the number $n$ of components of the order parameter and the dimension $d$ of the system. $n$ and $d$ are independent: one can define the Ising model ($n = 1$) in $d = 3$, 2 or 1 dimensions. Systems with the same $(n,d)$ values have the same behavior and the same exponents, irrespective of the specific interactions and local structure. This universality is the beauty of critical phenomena.

We describe in the next section a hundred years of theoretical models developed to describe these features and to calculate the critical exponents.

### References

1. H. Eugene Stanley, *Introduction to Phase Transitions and Critical Phenomena*. Oxford, UK: Clarendon Press.
2. P.M. Chaikin, T.C. Lubenski, *Principles of Condensed Matter Physics*. Cambridge University Press.
3. P.-G. de Gennes, Leçon Inaugurale au Collège de France. November 10, 1971.
4. P.-G. de Gennes, *Solid State Communications*, 10(9), 753–756 (1972).

## 3.3 MODELS OF PHASE TRANSITION: STATICS

We first describe the mean field theories, which allow us to understand qualitatively the features of critical phenomena but fail to give the right value of the critical exponent.

We then introduce the scaling laws and the renormalization theory of Wilson, which yield an exact calculation of critical exponent versus two variables only, $n$ and $d$.

### 3.3.1 Phenomenological Theories

#### 3.3.1.1 Molecular Field

In 1907, P. Weiss gave the first statistical description of ferromagnetism, introducing the notion of a molecular field [1]. A similar concept already existed in van der Waals' work on the equation of state of simple liquids (1873). In a paramagnetic material, the magnetic dipoles of the different atoms are independent. A magnetization of the medium appears if a magnetic field is applied, and it is given by Curie's law:

$$M = \frac{N\mu^2 H}{k_B T},\tag{3.4}$$

where $N$ is the density of magnetic dipoles $\mu$.

If the magnetic dipoles of the different atoms are now coupled by interactions that promote parallel alignment, Weiss assumed that each moment is not only subjected to the external field but to a field created by its neighbors that will be larger for a higher alignment [1]. The molecular field is proportional to the magnetization $M$: $H_m = \lambda M$.

Inserting $H + H_m$ in Equation 3.4 leads to:

$M = (N\mu^2/k_B T)(H + \lambda M)$, which can also be written:

$$M/N\mu = \mu/k_B (T - T_c) \quad \text{with} \quad k_B T_c = \lambda N \mu^2$$

*Numerical application*: calculate $\lambda$ for iron $T_c = 1{,}000$ K; $\mu = 9.2 \times 10^{-24}$ J T$^{-1}$, $N = 9 \times 10^{28}$ atoms m$^{-3}$.

Other models were proposed, but the general and unified version was introduced by the great Soviet physicist L.D. Landau in 1937. Mean-field theories are mathematically simple and allow the analytical calculation of phase diagrams and critical properties.

### 3.3.1.2 Landau Theory: Second-Order Magnetic Transition

Landau theory is based on a power series expansion of the free energy in terms of the order parameter $M$ [2]. It is convenient to define an adimensional parameter $m = M/M_s$, where $M_s = N\mu$ is the magnetization at saturation.

$$F\big|_{m^3} = \frac{1}{2} r_0 m^2 + \frac{1}{4} u_0 m^4 + \frac{1}{2} K (\nabla m)^2$$

where $r_0 = r_0^* (T - T_c)$ is small and vanishes at $T_c$; $u_0 \sim k_B T/a^3$ and $K \sim k_B T/a^3$ are regular at the transition. A characteristic length $\xi$ is defined by $\xi^2 = K/r_0$. Notice that $F$ is an even function of $M$ because $F(M) = F(-M)$. When a magnetic field is applied, we define the free enthalpy:

$$G\big|_{m^3} = \frac{1}{2} r_0 m^2 + \frac{1}{4} u_0 m^4 + \frac{1}{2} K (\nabla m)^2 - m M_s H$$

The plots of $F(M)$ and $G(M)$ at different temperatures are shown Figure 3.10.

*Order parameter $M_{eq}$*

$(\partial F/\partial M) = 0$ leads to $r_0^* (T - T_c) m + u_0 m^3 = 0$

and $M = M_S \sqrt{r_0^*/u_0 (T - T_c)}$, i.e. $M \sim (-\varepsilon)^{1/2}$

The Landau exponent $\beta$ is $\beta = 0.5$.

*State equation $H = (\partial F/\partial M)\big|_T$*

The derivative of $F$ with respect to $M$ leads to the isotherms $H(M)$ shown in Figure 3.10C.

$$H = r_0^* (T - T_c) \frac{M}{M_s} + u_0 \left(\frac{M}{M_s}\right)^3$$

*Magnetic susceptibility $\chi$*

By definition, $\chi^{-1} = (\partial H/\partial M)\big|_T = (r_0^*/M_s)(T - T_c) + 3(u_0/M_s^2) M^2$

It leads to:

$T > T_C$: $\chi = (M_s/r_0^*)(T - T_c)^{-1}$, i.e. $\chi = \varepsilon^{-1}$

$T < T_C$: $\chi = (M_s/2r_0^*)(T_c - T)^{-1}$, i.e. $\chi = (-\varepsilon)^{-1}$

The Landau exponent $\gamma$ is $\gamma = 1$. Below $T_C$, $G(M)$ has two minima separated by an energy barrier leading to hysteresis as shown in Figure 3.10A–C.

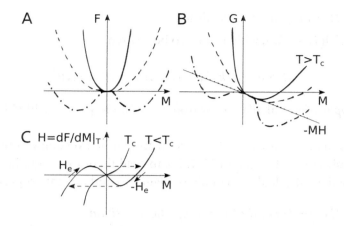

FIGURE 3.10  (A) Free energy density $F$ as a function of order parameter $M$ near the critical point for temperature above and below the critical temperature; (B) Free enthalpy density as a function of order parameter in constant magnetic field $H$; (C) $H$–$M$ isotherms ($H$ is the derivative of $F$ at constant $T$) showing hysteresis for $T < T_c$. $H_e$ is the coercive field.

*Thermal fluctuations of M*

It is convenient to analyze $M$ (or the adimensional variable $m$) in Fourier components $m_q$ defined by:

$$m_q = \int m(r)e^{i\vec{q}.\vec{r}}d_3\vec{r}$$

In the Fourier space, by using the quadratic approximation, $F$ becomes:

$$F \cong \sum_q \frac{1}{2}r_0 m_q^2 + \frac{1}{2}Kq^2 m_q^2 \text{ for } T > T_c$$

$$F \cong \sum_q -r_0 m_q^2 + \frac{1}{2}Kq^2 m_q^2 \text{ for } T < T_c$$

We can now derive the thermal average of $M_q^2$ from the equipartition theorem (stating that the energy is $k_b T/2$ per mode). It leads to the Ornstein–Zernike formula:

$$M_q^2 = M_s^2 \frac{k_B T}{r_0^*(T-T_c) + Kq^2} = \frac{M_s^2}{r_0^*(T-T_c)}\frac{1}{1+q^2\xi^2} \text{ for } T > T_c$$

$$M_q^2 = M_s^2 \frac{k_B T}{2r_0^*(T_c - T) + Kq^2} \text{ for } T < T_c$$

The magnetic correlation function $g(r)$ is defined by:

$$g(r) = \langle m(r)m(0)\rangle - m^2$$

$m_q^2$ is the Fourier transform of $g(r)$. It leads to:

$$g(r) \sim \frac{e^{-r/\xi}}{r}$$

where $\xi$ is the correlation length which diverges at $T_c$:

$$\xi = \sqrt{\frac{K}{r_0^* |T - T_c|}}$$

The Landau exponent $\nu$ is $\nu = \frac{1}{2}$.

From the fluctuation dissipation theorem, we can express the susceptibility $\chi$ as a function of $g$:

$$\chi = \frac{1}{k_B T} \int g(r) d_3 r = \frac{1}{k_B T} M_q^2(0)$$

$$\text{For } T > T_c, \ \chi = \frac{1}{r_0^* (T - T_c)}$$

$$\text{For } T < T_c, \ \chi = \frac{1}{2 r_0^* (T_c - T)}.$$

To summarize, the Landau theory explains with simple laws the main features of critical phenomena. Because the variation of the free energy as a function of the order parameter $F(M)$ is extremely flat at the critical temperature ($F' = F'' = F''' = 0$), the fluctuations are gigantic and the response to small perturbation is huge. As seen in Chapter 2, large responses to small perturbations are a characteristic of soft matter.

### 3.3.1.3 Landau–de Gennes Theory: First-Order Isotropic-Nematic Transition

The Landau theory had been extended to liquid crystals by Pierre-Gilles de Gennes in 1972 [3].

Near the transition, the free energy can be expanded as a function of the scalar nematic order parameter $S$:

$$F\big|_{m^3} = \frac{1}{2} r_0 S^2 - \frac{1}{3} w_0 S^3 + \frac{1}{4} u_0 S^4$$

where $r_0 = r_0^* (T - T_0)$. Note that $T_0$ is not the transition temperature $T_c$. The third order term is due to the fact that $F(S) \neq F(-S)$.

$S > 0$ corresponds to molecules that align in the direction of the director $\vec{n}$, whereas $S < 0$ corresponds to molecules that align perpendicular to $\vec{n}$.

The free energy $F(S)$ is shown in Figure 3.11A for different temperatures. The equilibrium value of $S$ corresponds to the minimum of the free energy $F$ for a given temperature. The evolution of the equilibrium value of $S$ as a function of $T$ is shown in Figure 3.11B. The transition is first order (the order parameter is discontinuous at the transition).

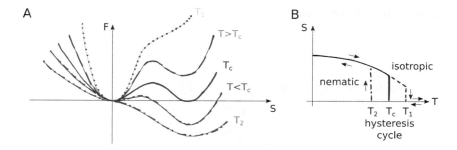

FIGURE 3.11 **(See color insert.)** (A) Free energy isotherms as a function of $S$; (B) Order parameter $S$ as a function of $T$, showing a first-order transition.

The critical temperature $T_c$ is derived from the two equations:

$$\frac{\partial F}{\partial S} = 0, \text{ i.e. } r_0 - w_0 S_C + u_0 S_C^2 = 0$$

$$F = 0, \text{ i.e. } \frac{1}{2}r_0 - \frac{1}{3}w_0 S_C + \frac{1}{4}u_0 S_C^2 = 0$$

Leading to $S_C = 2w_0/3u_0$ and $T_c = T_0 + \left(2w_0^2/9u_0 r_0^*\right)$.

A general phenomenon associated with first-order phase transitions is the hysteresis in cycling trough the transition, leading to superheated and supercooled phases. This is shown in Figure 3.11B. The hysteresis comes from the fact that the system is blocked in a minimum energy (metastable state) and only transits to another state if this local minimum disappears.

For $T_2 < T < T_1$ (Figure 3.11): coexistence between isotropic and nematic phases separated by an energy barrier. In practice, inclusions of the nematic phase appear in the isotropic phase (or vice versa), with a clear interface between the two phases.

### 3.3.2 Scaling Laws: Renormalization Theory

Molecular field failures appear both theoretically and experimentally for the determination of critical exponents as shown in Tables 3.2 and 3.3. Onsager in 1944 found an exact solution to the two-dimensional Ising model ($n = 1$, $d = 2$). The susceptibility and the correlation length do not follow Landau's exponents at all. From 1958 onwards, new techniques

TABLE 3.2   Values of Critical Exponents for Selected Physical Systems [4]

|  | Xe | Bin.mixt. | β-brass | $^4$He | Fe | Ni |
|---|---|---|---|---|---|---|
| $n$ | 1 | 1 | 1 | 2 | 3 | 3 |
| β | $0.35 \pm 0.15$ | $0.322 \pm 0.002$ | $0.305 \pm 0.005$ |  | $0.37 \pm 0.01$ | $0.358 \pm 0.003$ |
| γ | $1.3 \pm 0.2$ | $1.239 \pm 0.002$ | $1.25 \pm 0.02$ |  | $1.33 \pm 0.015$ | $1.33 \pm 0.02$ |
| δ | $4.2 \pm 0.6$ |  |  |  | $4.3 \pm 1$ | $4.29 \pm 0.05$ |

TABLE 3.3    Critical Exponents for Different Models as a Function of the Dimensionality $n$ of the Order Parameter and the Dimensionality $d$ of Space [4]

|  | Mean-field | Ising $d=2$ | Ising $d=3$ | Heisenberg | Spherical |
|---|---|---|---|---|---|
| $(n, d)$ |  | $(1, 2)$ | $(1.3)$ | $(3, 3)$ | $(\infty, 3)$ |
| $\beta$ | 1/2 | 1/8 | $0.326 \pm 0.004$ | $0.38 \pm 0.03$ | 1/2 |
| $\gamma$ | 1 | 7/4 | $1.239 \pm 0.003$ | $1.38 \pm 0.02$ | 2 |
| $\delta$ | 3 | $(15)$ | $4.80 \pm 0.05$ | $4.65 \pm 0.29$ | 5 |

such as nuclear magnetic resonance, light and neutron scattering have allowed the accurate measurement of critical exponents [4].

### 3.3.2.1 Breakdown of Mean Field Theory: Ginsburg Criterion

The critical exponents are not given in general by the Landau theory as shown in Table 3.2. Mean field theory replaces a *fluctuating local order parameter* with a spatially uniform *average order parameter*. This is a good approximation if the fluctuations of the order parameter are small compared to its mean value.

A quantitative criterion was first introduced by V.L. Ginsburg in 1960 [5]. In a $d$-dimensional space, the amplitude of the fluctuations of the magnetization $M$ can be estimated by considering that the energy of a fluctuation of $M$ in a characteristic volume $\zeta^d$ is the thermal energy $k_B T$:

$$\left\langle \delta M^2 \right\rangle r_0 \xi^d \sim k_B T$$

$$\frac{u_0 \delta M^4}{r_0 \delta M^2} = \frac{u_0}{r_0^2} \frac{k_B T}{\xi^d} \sim r_0^{d/2-2} \sim \varepsilon^{d/2-2} \ll 1 \; if \; d > 4.$$

The fluctuation can then not be neglected in the usual case $d = 3$.

### 3.3.2.2 Universality and Similarity: Scaling Laws

In the 1960s, physicists were confused by the flowering of critical exponents. Fortunately, two empirical laws have emerged from the results accumulated from calculations and experiments.

The first law is the *universality* of critical phenomena: critical exponents depend only on $n$ and $d$. The Ising model applies to the liquid–gas transition or to the phase separation between two chemical species. This is the beauty of the field.

The second law is *similarity*. B. Widom and L. Kadanoff [6, 7] introduced the notions of thermodynamic similarity and spatial similarity respectively: near the critical point a change in temperature is equivalent to a change in the measurement scales for magnetizations, correlation lengths, and all the physical variables associated with giant fluctuations; if you take a picture of the system at two different temperatures near $T_c$, the scales of the fluctuations are different but the images are indistinguishable if you make a change in scale. The resulting scaling laws establish relationships between critical exponents. It allows the expression of the usual exponents with only two of them.

However, these laws do not allow the calculation of critical exponents. In 1972, K.G. Wilson at Princeton devised a method to calculate them.

### 3.3.2.3 Renormalization Wilson Theory

The Wilson method [8] is based on a change of scale and schematically explained in Figure 3.12.

Instead of discussing the statistics of individual magnetic moments (thin arrows), they are first grouped into larger blocks. The resulting moments (thick arrows) follow the same laws as the individual moments if we change the temperature scale. Wilson thus showed that this method allows the calculation of all the critical exponents of a given system.

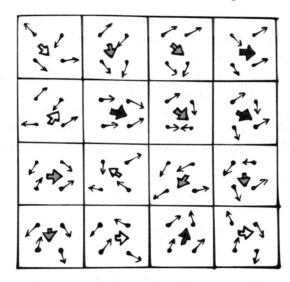

FIGURE 3.12    Illustration of the spin blocks.

As explained by the Ginsburg criterion, the fluctuations are too large and the series expansion versus the order parameter is not convergent, except in a space of dimension $d > 4$. This led K. Wilson and M. Fisher to study the case of dimensions slightly less than 4, $d = 4 - \varepsilon$, where $\varepsilon$ is a small parameter. In 1972, Wilson and Fisher could calculate the critical exponents as series in $\varepsilon = 4 - d$. At zero order, Landau's exponents were found as expected, and by setting $\varepsilon = 1$, we can investigate the realistic case of three-dimensional systems. The results are numerically remarkable in 3-D systems since they give all known critical exponents within a few percent. Table 3.4 shows the Wilson exponent (renormalization group (RG)) for various values of $n$, the dimension of the order parameter.

TABLE 3.4    Critical Exponents from Three-Dimensional
Renormalization Group (RG) and Series. Adapted from [9]

| $n$ | Exponent | RG | e-expansion | Series |
|-----|----------|-----|-------------|--------|
| $n = 0$ | $\beta$ | $0.302 \pm 0.0015$ | 0.305 | 1.615–1.617 |
| | $\gamma$ | $1.1615 \pm 0.0020$ | 1.163 | 0.60 |
| | $\nu$ | $0.588 \pm 0.0015$ | 0.589 | |
| $n = 1$ | $\beta$ | $0.325 \pm 0.0015$ | 0.330 | 0.303–0.318 |
| | $\gamma$ | $1.241 \pm 0.00020$ | 1.242 | 1.241–1.250 |
| | $\nu$ | $0.630 \pm 0.0020$ | 0.632 | 0.638 |
| $n = 2$ | $\beta$ | $0.3455 \pm 0.0020$ | 0.357 | |
| | $\gamma$ | $1.316 \pm 0.0025$ | 1.324 | |
| | $\nu$ | $0.669 \pm 0.0020$ | 0.676 | |
| $n = 3$ | $\beta$ | $0.3645 \pm 0.0025$ | 0.379 | 1.315–1.333 |
| | $\gamma$ | $1.386 \pm 0.0040$ | 1.395 | 0.670–0.678 |
| | $\nu$ | $0.705 \pm 0.0030$ | 0.713 | |

Wilson's theory was extended by Pierre-Gilles de Gennes to polymer physics by setting $n = 0$. The arsenal of techniques developed for critical phenomena had completely changed the physics of polymers, in particular the use of scaling laws.

References

1. P. Weiss, *J. Phys. Radium*, 6, 667 (1907).
2. L.D. Landau, E.M. Lifshitz, *Statistical Physics*, 3rd Edition. Elsevier, 1980.
3. P.-G. de Gennes, *The Physics of Liquid Crystals*. Oxford University Press, 1974.
4. H.E. Stanley, *Introduction to Phase Transitions and Critical Phenomena*. Oxford University Press, 1971.
5. V.L. Ginzburg, *Fiz. Tverd. Tela*, 2, 2031 (1960).
6. L.P. Kadanoff, *Phys. Physique Fiz.*, 2, 263 (1966).
7. B. Widom, *J. Chem. Phys.*, 43, 3898 (1965).
8. K.G. Wilson, J. Kogut, *Phys. Rep.*, 12, 75–199 (1974).
9. J.C. Le Guillou, J. Zinn-Justin, *Phys. Rev. Lett.*, 39, 95 (1977).

## 3.4 DYNAMICS OF CRITICAL PHENOMENA: "Z" EXPONENT

When a system is close to a critical point, anomalies occur in static and dynamical properties. The fluctuations of the order parameter $\psi$ become giant and characterized by a correlation length $\xi$ that diverges at $T_c$. Their dynamics slow down, and the characteristic time $\tau_\psi$ diverges at $T_c$. This phenomenon is called *critical slowing down*. It is due to the shape of the thermodynamic potential $F(\psi)$, which becomes very flat at the critical temperature, leading to very low forces. It illustrates the *albatross theorem* "what is big is slow" introduced by Pierre-Gilles de Gennes for the dynamics of macromolecules, in reference to Baudelaire's poem "*Ses ailes de géant l'empêchent de marcher*" (His giant wings prevent him from walking).

The dynamical properties may be measured by inelastic light or neutron scattering experiments, which lead to the time-dependent correlation functions and relaxation rates of the fluctuations. The description of dynamic properties is more complex than the statics, because it's less universal. For example, the liquid–gas transition and the magnetic Ising model ($n = 1$, $d = 3$) have the same critical exponents, but the dynamical laws are completely different because in one case $\psi$ is conserved and in the magnetic case it is not.

However, we shall see that the dynamic properties of a system close to or at the critical point are characterized by one exponent "z," named the *dynamical exponent*.

Hydrodynamic theories apply at lengths smaller than the correlation length $\zeta$ (or wave vector $q \xi < 1$) and times larger than the characteristic time $\tau_\psi$, which both diverge at $T_c$. From the calculation relaxation times $\tau_\psi(q)$ of the collective mode of wave vector $q$, in the hydrodynamic limit in the phase below or above $T_c$, one can extend the results to the critical regime using *dynamical scaling laws*. In particular, the dependence upon the correlation length $\zeta$ disappears in the limit $q \xi > 1$. In addition, even in the hydrodynamic regime $q \xi < 1$ and $t > \tau_\psi$, the fluctuations of the order parameter give rise to singularities of certain reactive and dissipative parameters.

To describe the dynamics of critical phenomena, three types of approaches have been developed:

i) Van Hove approximation

ii) Dynamical scaling laws (Halperin, Hohenberg)

iii) Mode–mode coupling (Kawasaki, Ferrell, Kubo)

We will briefly present these three types of descriptions, which account fairly well for the dynamical properties of superfluid, magnetic systems, binary mixtures, and liquids.

### 3.4.1 Van Hove Approximation

Historically, the dynamic properties of phase transitions were described by the "conventional theory" or Van Hove approximation [1].

#### 3.4.1.1 Non-Conserved Order Parameter (Magnetization, Nematic Order, etc.)

If the order parameter $\psi$ is not conserved, the relaxation of $\psi$ is described by a time-dependent Landau equation:

$$\gamma_v \frac{\partial \psi}{\partial t} = -\frac{\partial}{\partial \psi} \tag{3.5}$$

where $-(\partial F/\partial \psi)$ is the conjugated force to $\psi$.

For small $\psi$, we can write $F = (1/2)a\psi^2 = (1/2\chi)\psi^2$, where $\chi$ is the susceptibility that diverges at $T_c$. The time constant $\tau_\psi$ associated to the relaxation of $\psi$ is:

$$\tau_\psi = \gamma_v \chi$$

Van Hove assumes that the coefficient $\gamma_v$ remains regular in temperature because it reflects microscopic frictions that depend only on short range correlations. It leads to a divergence of $\tau_\psi$ for $T \to T_c$ with the same exponent as $\chi$. In mean-field theory, $\tau_\psi \sim \varepsilon^{-1}$

This divergence of $\tau_\psi$ is referred to as "critical slowing down."

### 3.4.1.2 Conserved *Order Parameter (Binary Mixture, Liquid–Gas, etc.)*

If $\psi$ is conserved, Equation 3.5 is replaced with:

$$C\frac{\partial \psi}{\partial t} = \nabla^2\left(-\frac{\partial F}{\partial \psi}\right) = r_0\nabla^2\psi \tag{3.6}$$

where $C$ is a friction coefficient.

$\psi$ obeys a diffusion equation, with a diffusion coefficient $D = r_0/C$. The relaxation time of a fluctuation of wave vector $q$ is:

$$\tau_\psi(q) = \frac{1}{Dq^2}$$

In the Van Hove approximation, we assume $C$ regular and $D \sim r_0 \sim T - T_C$.

To illustrate the physics behind Equation 3.6, we apply the Van Hove approximation to describe the dynamics of a binary mixture near the critical point.

Consider a fluctuation of concentration $c$ of a solute A in a solvent B, shown in. It is convenient to examine a weak, sinusoidal modulation of $c$, as shown in Figure 3.13.

$$c(x,t) = c + \delta c_q(t)\cos(qx).$$

In the $c(x)$ profile, we see high density region $(M,M')$ and low density region $(P,P')$. There is an elastic force $f$ from $M$ to $P$ tending to equalize the concentrations: this fluctuation of concentration induces a modulation of the chemical potential $\mu$, which gives rise to a flow:

$$\varsigma_0 v = -\frac{\partial \mu}{\partial x} \tag{3.7}$$

where $\varsigma_0 = \eta a$ is friction coefficient for the molecule A of size $a$ and $\eta$ the viscosity of the solvent. Equation 3.7 describes the balance between the friction force $\varsigma_0 v$ and the driving force $-(\partial \mu/\partial x)$ per molecule.

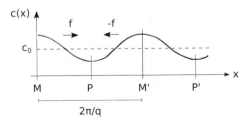

FIGURE 3.13 Fluctuation of concentration of molecules A in an A/B mixture of wave vector $q$.

The flux of A is $J = cv$.

$$J = \frac{c}{\varsigma_0}\frac{\partial \mu}{\partial x} = -\frac{c}{\varsigma_0}\frac{\partial \mu}{\partial \Pi}\frac{\partial \Pi}{\partial c}\frac{\partial c}{\partial x} = -D\frac{\partial c}{\partial x} \tag{3.8}$$

where $\Pi$ is the osmotic pressure.

From the thermodynamic relationship $\partial \mu / \partial \Pi = 1/c$ and the definition of the osmotic compressibility $\chi = \partial c / \partial \mu$, one finds the diffusion coefficient:

$$D = \frac{1}{\varsigma_0 \chi} = \frac{k_B}{\eta a}(T - T_c) \tag{3.9}$$

By adding the conservation equation: $(\partial c / \partial t) + (\partial J / \partial x) = 0$, we obtain the characteristic time of the modes as a function of $T_c$:

$$\frac{1}{\tau_\psi(q)} = Dq^2 \sim T - T_c$$

The Van Hove approximation is the extension for dynamics of the Landau theory for statics. It describes some transitions quite well, as for example smectic A/smectic C or the dynamics of the second order Frederick's transition in a nematic liquid crystal induced by a magnetic field, for which $\varepsilon = (H - H_C)/H_C$ [2]. But for many transitions, such as phase separation in binary mixtures, the Van Hove approximation is incorrect; it has to be replaced by a less restrictive assumption. The most powerful technique is the hypothesis of dynamical similarity that we will now recall.

## 3.4.2 Dynamical Scaling Laws

### 3.4.2.1 Universality of Critical Exponents

The critical "static" and "dynamic" exponents which characterize the divergence of certain quantities at the critical point are "universal," in that they do not depend on microscopic interactions [3, 4].

- For the static properties, the critical exponents depend only on $n$, the order parameter dimension, and $d$, the space dimension. The various exponents are bound by linear equations, and it is sufficient to know two of them to calculate all the others. These relationships are direct consequences of static scaling laws based on two assumptions: (1) the thermodynamic similarity hypothesis, which postulates homogeneity for thermodynamic functions leading to power laws, and (2) the spatial similarity hypothesis, based on the existence of one single length $\xi(T)$ which describes the correlations of the order parameter.

- For the dynamic properties, the knowledge of $n$ and $d$ is not sufficient. Critical behavior depends on symmetry and conservation laws but remains independent of the details of microscopic interactions. A principle of universality more limited still remains; thus, superfluid helium and an isotropic magnetic transition have identical dynamic properties, while an Ising transition has a completely different dynamic behavior. This complexity of dynamic behavior explains why very powerful methods developed by the renormalization group for static properties have only led to a few results for dynamic properties. The tool of the most fruitful work remains that of dynamic scale laws.

### 3.4.2.2 Statement of Dynamic Scaling Laws

For the dynamic properties, a double hypothesis of spatial similarity (a single relevant length) and temporal similarity (a relevant time) is made.

The dynamic and static correlation functions are defined by:

$$g(r,t) = \left\langle \psi(r,t) - \left\langle \psi(r,t) \right\rangle \right\rangle \left\langle \psi(0,0) - \left\langle \psi(0,0) \right\rangle \right\rangle$$

$$g(r,t) = \int \frac{d_3 q}{(2\pi)^3} \int \frac{d\omega}{2\pi} e^{i(\vec{q}\cdot\vec{r} - \omega t)} g(q,\omega)$$

$g(q,\omega)$ can be written as:

$$g(q,\omega) = \frac{g(q)}{\omega(q)} f\left(\frac{\omega}{\omega(q)}\right)$$

The static scaling laws assume that $g(q)$ is a homogeneous function of $q$ and $\xi$:

$$g(q) = q^x G(q\xi) \tag{3.10}$$

The dynamic scale laws assume that the characteristic frequency $\omega(q)$ is also a homogeneous function of $q$ and $\xi$ (Figure 3.14:

$$\omega(q) = q^z \Omega(q\xi) \tag{3.11}$$

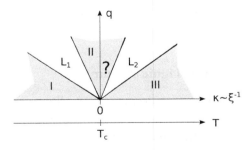

FIGURE 3.14 The three hatched regions I, II, III of the plane $q$, $\xi^{-1}$ are defined respectively by the conditions $(q\xi \ll 1, T < T_c)$, $(q\xi \gg 1, T = T_c)$, $(q\xi \ll 1, T > T_c)$.

In the three regions I, II, III shown in Figure 3.14, the spatial temporal correlation functions have different behaviors: the scaling laws assume that the asymptotic forms extrapolated to lines $L_1$ and $L_2$ ($q\xi = 1$, $T < T_c$ and $T > T_c$) must merge. I and III represent the validity domain of hydrodynamic theories, which allow the calculation of the relaxation rates.

$\Omega$ and $f$ being homogeneous functions, one can deduce the behavior in one region (for example II in Figure 3.14) from the properties in another region (for example I or III in Figure 3.14). From the derivation of $\omega(q)$ in Regions I and III, the dynamical scaling hypothesis (Equation 3.11) condition imposes:

$$\omega^I\left(\xi^{-1}\right) = b_1\omega^{II}\left(\xi^{-1}\right)$$

$$\omega^{III}\left(\xi^{-1}\right) = b_3\omega^{II}\left(\xi^{-1}\right)$$

where $b_1$ and $b_3$ are numerical constants of order unity.

### 3.4.2.3 Application of Dynamical Scaling Laws

Dynamic scaling laws predict the behavior of the fluctuations of the order parameter in the Regions II or III if it is known in Region I (in ordered phase) by hydrodynamic arguments. When the corresponding mode is propagative (low damping), its behavior is only related to thermodynamic parameters, and it is possible to express the dynamical exponent "z" in terms of static exponents. This holds for superfluid helium, Heisenberg ferromagnetism, smectic A-nematic, etc. For these systems, there is a continuous broken symmetry in the ordered phase (e.g. rotation of the magnetic moments, translation in the normal direction to smectic layers). One can define "quasi-conserved" variables, the *phase* in a superfluid, the *direction of magnetization* in a ferromagnetic, the *director n* in a nematic, the *displacement u* of the layers in a smectic, that enter the hydrodynamic equations and give rise to collective modes (second sound for helium, spin waves, second sound of smectics A). The frequency of the modes is determined in Region I of Figure 3.14. It involves a constant rigidity, which depends upon temperature with a known exponent. The scaling laws allow the extension of the results to other domains.

We will illustrate this approach for two cases: (1) the lambda transition of superfluid helium [4] similar to the case of the smectic A–nematic transition in liquid crystals [5], and (2) the liquid–gas transition

**Example 1: Lambda Transition in Liquid Helium**

The order parameter is the wave function $\psi = \rho_s\, e^{i\phi}$, where $\rho_s$ is the superfluid density and $\phi$ the phase. The superfluid velocity $v_s$ is proportional to the gradient of $\phi$.

The dynamic correlation function of the order parameter is dominated by the second sound, whose frequency is:

$$\omega(q) = \pm c_2 q + i D_2 q^2 \tag{3.12}$$

where $c_2{}^2 \sim \rho_s \sim \xi^{-1}$.

Above $T_c$, the fluctuations of $\psi$ relax in a finite time $\tau(T)$.

The dispersion relationship of the second sound allows us to determine the dynamical exponent $z$ and $\tau(T)$:

$$\omega(q) \sim \xi^{-1/2} q \sim q^{3/2} \Omega(q\xi), \text{ i.e } z = \frac{3}{2}$$

The dynamical scaling laws give

i)  the characteristic frequency in the critical region:

$$\omega(q, T_c) = A q^{3/2}$$

ii)  the damping of the second sound:

$$D_2 \sim c_2 \xi \sim \xi^{1/2}$$

iii)  the characteristic time of the fluctuations of $\psi$ above $T_c$:

$$\tau_\psi(T) \sim \xi^{3/2}$$

The same approach allows the description of the dynamic behavior of the nematic–smectic A transition [5], where second sound modes in the smectic phase slow down at the transition and become diffusive in the nematic phase.

**Example 2: Liquid–Gas Transition**

For the liquid–gas transition at the critical point, the spectrum of density fluctuations is dominated by the thermal diffusion mode $\omega = D_T q^2$.

The dynamic laws of scale only allow us to conclude that $a = a'$ and $z = 2 - a$, but we cannot derive $a$. We shall see that mode–mode coupling approaches will allow the derivation of "$z$."

## 3.4.3 Linear Response Theory

The linear response theory leading to the fluctuation dissipation theorem states that the linear response of a given system to an external perturbation is expressed in terms of the fluctuations of the system in thermal equilibrium. This approach leads to the mode–mode coupling theory, which has been applied with great success to binary mixtures near the critical point by Kawasaki and Ferrell [6]. Without any reference to the properties of the low temperature phase, anomalies of transport coefficients are determined due to the critical fluctuations of the order parameter in the high temperature phase. Correlated domains of size $\xi(T)$ appear momentarily (life time $\tau(T)$) with the properties of the ordered phase. Kubo's formulas [7] establish relationships between transport coefficients and correlation function of fluxes, which allow the interpretation of anomalies of the transport coefficient which diverge at the critical temperature.

This approach had been used to calculate the diffusion coefficient $D$ describing the relaxation of critical fluctuations of a binary mixture, known as the Kawasaki–Einstein–Stokes formula [8].

We have seen that a modulation of the concentration, shown in Figure 3.13, induces a flow $J$, which can be written as:

$$v = -s\nabla\mu$$

$$J = -cs\frac{\partial\mu}{\partial\Pi}\frac{\partial\Pi}{\partial c}\nabla c = -\frac{s}{\chi}\nabla c = -D\nabla c \tag{3.13}$$

where $s$ is the mobility. In the Van Hove approximation, $s = (\eta_0 a)^{-1}$ is approximated by a constant. It neglects the hydrodynamic correlations between the solute molecules. Including the back-flows due to surrounding molecules leads to:

$$s = \int \frac{1}{6\pi\eta r}g(r)dr$$

where $\eta$ is the solvent viscosity.

It is also possible to relate the osmotic compressibility $\chi = \frac{\partial c}{\partial\Pi}$ with the pair correlation function $g(r)$:

$$\chi = \frac{1}{k_B T}\int g(r)dr$$

Using the Orstein–Zernike formula $g(r) \cong \left(1/a^2 r\right)e^{-r/\xi}$ and the definition of $D$ (Equation 3.13) leads to:

$$D = \frac{k_B T\int \frac{1}{6\pi\eta r}g(r)dr}{\int g(r)dr} = \frac{k_B T}{\eta\xi} \tag{3.14}$$

The frequency spectrum of the critical fluctuation is given by:

$$\omega(q) = \frac{k_B T}{\eta\xi}q^2 = q^3 \times \frac{k_B T}{\eta(q\xi)}$$

This leads to a dynamical exponent $z = 3$. In the critical region II Figure 3.14, $q\xi > 1$, $\omega q \sim q^3$. The droplets of size $\xi$ in the monophasic region relax with a time $\tau \sim \eta\xi^3 / k_B T$.

The Kawasaki formula $D = k_b T/6\pi\eta\xi$ was experimentally verified by P. Bergé and M. Dubois [9] as shown Figure 3.15A. They measured $D$ from the spectral width of light scattered by concentration fluctuations of cyclohexane–aniline critical mixture near $T_c$. It leads to a determination of $\nu = 0.60\pm0.01$ (Figure 3.15B).

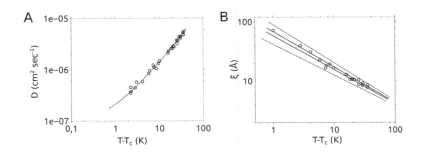

FIGURE 3.15 (A) Variation of the diffusion coefficient vs. T – $T_C$:triangle, measured coefficient; circle, calculated from Equation 3.14. (B) Variation of ξ vs. $T$ – $T_C$. (Adapted from [9].)

We will see in Section 7.7.2.2 that Equation 3.10 holds also for polymer solutions: $D_{coop} = k_B T/\eta \xi$, where ξ is the mesh size of the semi-dilute solution, and the dynamical exponent for polymer solution is $z = 3$.

## References

1. L. Van Hove, *Phys. Rev.*, 93, 268 (1954).
2. P. Pieranski, F. Brochard, E. Guyon, *J. Phys.*, 34, 35 (1973).
3. B.I. Halperin, P.C. Hohenberg, S.-K. Ma, *Phys. Rev. Lett.*, 29, 1548 (1972).
4. R. Perl, R.A. Ferrell, *Phys. Rev. Lett.*, 29, 51 (1972).
5. F. Brochard-Wyart, *Phys. Lett.*, 49, 315 (1974).
6. R.A. Ferrell N. Menyhard, H. Schmidt, F. Schwabl, P. Swepfalusy *Ann. Phys.*, 47, 565 (1968).
7. R. Kubo, *J. Phys. Soc. Jpn*, 12, 570 (1957).
8. A.F. Kholodenko, J.F. Douglas, *Phys. Rev. E*, 51, 1081 (1995).
9. P. Bergé, M. Dubois, *Phys. Rev. Lett.*, 27, 1125 (1971).

# CHAPTER 4

# Interfaces

## 4.1 COLLOIDAL SYSTEMS: CLASSIFICATION AND FABRICATION

### 4.1.1 General Features

Colloidal systems are ubiquitous in our everyday life. We will see numerous examples in the fields of food industry and agriculture; they are also key in the formulation of paints and cosmetics.

The colloidal state is characterized by:

- The dispersion of substances (molecules, macromolecules, particles) in a continuous medium (gas, liquid, or solid).
- The presence of mesoscopic size "discontinuities" (in the 1–100 nm range – Section 2.1), which may originate from the dispersed phase (particle size) or from the continuous phase (pore size).

These two criteria imply that colloidal systems have a high surface area-to-volume ratio S/V. For the sake of clarity, crushing a cubic piece of 1 cm side into small 10 nm particles leads to a one million-fold increase of S/V, and the particles will cover the surface of two tennis courts. The properties of these divided systems will therefore be essentially controlled by the physicochemical phenomena and processes occurring at the interfaces between the dispersed phase and the continuous phase. We have seen the importance of the electrostatic double layer (Section 2.4), which guides the stabilization or flocculation of colloidal systems. Interfaces can also be affected by the adsorption of molecules. As a consequence, a trace amount of impurities dispersed in bulk with a high affinity for interfaces may be ultimately accumulated in an interfacial monomolecular layer and drastically alter the physical properties of the colloidal systems. Those molecules that have a high affinity for interfaces are generally called surfactants (Section 6.1) (Figure 4.1).

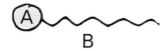

FIGURE 4.1 Surfactant molecule. The polar (hydrophilic) head A is attached to one (or two) hydrocarbon (hydrophobic) chain(s).

## 4.1.2 Classification

There are several ways to classify colloids. A distinction between lyophilic and lyophobic colloids is sometimes used. Here, we focus on a classification based on the "organizational" level of the colloidal system.

### 4.1.2.1 Stochastic Systems

Colloidal suspensions are composed of particles characterized by at least one dimension of sub-micrometric size. These particles are subjected to thermal agitation and are distributed randomly within the continuous phase. The nature (liquid or solid) and the shape (spherical, elongated) of the particles are variable (Figure 4.2).

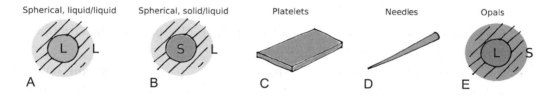

FIGURE 4.2 Different examples of colloidal suspensions. (A) *Spherical liquid/liquid*: Droplets of water in oil or of oil in water, milk, cosmetics, pharmaceutical or food products; (B) *Spherical solid/liquid*: carbon black, latex (paint); (C) *Platelets*: Clays (kaolin, paper coating); (D) *Needles*: cotton, paper; (E) *Opal*: inclusion of microdrops in silica.

Porous systems are part of the same category (Figure 4.3). They are solids containing interconnected voids (pores with a diameter between a few nm and a few μm). Examples include sponge, paper, meringue, sandstone, cement, and absorbents. They are also widely used in chromatography, heterogeneous catalysis, filtration, and ultrafiltration.

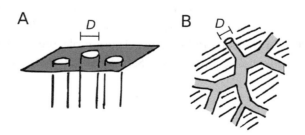

FIGURE 4.3 Porous systems. (A) Nuclear membrane (calibrated pores with diameter $D \sim 1$ to 100 nm). (B) Sponge, porous rock.

### 4.1.2.2 Self-Organized Systems

When not diluted with water, surfactants form smectics (e.g. soap). In aqueous solutions, surfactants form aggregates at low concentrations (e.g. micelles) and organized bicontinuous structures at high concentrations (e.g. cubic phase). Figure 4.4 summarizes the different possible structures of self-organized colloidal systems

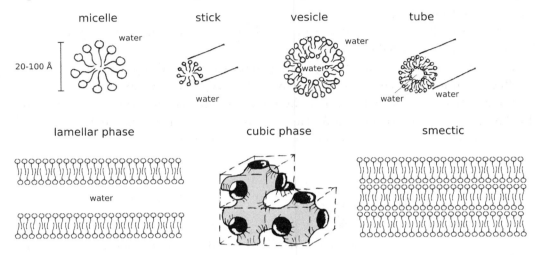

FIGURE 4.4   Different examples of self-organized colloidal systems.

### 4.1.2.3 Ternary Systems

Surfactants also play a role in making immiscible compounds compatible with their medium. For example, they serve to stabilize emulsions of water in oil or oil in water. Foams or soap films can be considered as a case of water-in-air mixing (Figure 4.5).

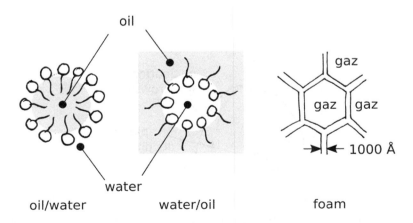

FIGURE 4.5   Role of surfactants in immiscible systems.

### 4.1.3 Preparation of Divided Systems

Colloidal particles of mesoscale size, i.e. intermediate between the macroscopic scale and the molecular scale, are manufactured either by degradation of the bulk material in a mechanical way or by aggregation of small molecules in a chemical way.

#### 4.1.3.1 Mechanical Way

The increase of the surface-to-volume ratio of the material is obtained by transforming the mechanical work supplied into surface energy.

For solids, the most direct way is grinding. In the case of liquids, agitation (Figure 4.6A) is effective in forming emulsions (e.g. mayonnaise, egg white) or foam. Another way of forming droplets is to form a stream of liquid that spontaneously breaks into small drops by Rayleigh instability (Figure 4.6B). Here, the mechanical energy is supplied during the injection of the liquid. This is how homogenized milk is made.

FIGURE 4.6  (A) Generation of an emulsion by stirring. (B) Generation of droplets through the Rayleigh instability of a liquid jet.

#### 4.1.3.2 Chemical Route

The formation of the dispersed phase takes place by precipitation of a binary mixture. Starting from a homogeneous phase where liquids A and B are miscible, if the temperature is suddenly lowered, a phase separation occurs by nucleation and growth or spinodal decomposition (Figure 4.7). Complex structures are formed and these can be freeze-dried.

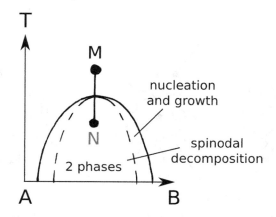

FIGURE 4.7  Phase diagram of a binary mixture. By lowering the temperature, one goes from a monophasic state (M) to a biphasic state (N). The solid line delineates the demixing zone. Outside this region, the two compounds A and B are miscible and form a single phase. The dotted line curve delimits the spinodal decomposition region for which the mixture is unstable and separates by amplification of concentration fluctuations.

## 4.2 CAPILLARITY AND SURFACE TENSION

Capillarity describes the ability of mobile interfaces (liquid–air or liquid–liquid) to deform to minimize their surface energy. It applies to phenomena observed in everyday life, like dew on blades of grass at dawn, rising sap in trees, imbibition of porous media, or the shape of raindrops on windshields. It also has many industrial applications: in cosmetics, food and painting industries, emulsions are produced. It is necessary to make stable emulsions while surface tension effects lead to the ingestion of small drops by bigger ones. In the coating of fibers in the textile industry or the formulation of mascara, it is important to find ways to cover threads and lashes with a regular and continuous sheath; understanding capillarity is to avoid the formation of a string of droplets.

In this section and the following ones on wetting, we will describe the basic theoretical concepts. More details on these phenomena can be found in references [1] and [2].

### 4.2.1 Surface Tension

The main property of liquids is that they flow. Yet they may adopt very stable geometric shapes. A soap bubble or a drop of oil in water form perfect spheres. The surface of a liquid behaves like a stretched membrane (Figure 4.8A), characterized by a surface tension. Vesicles, which are composed of insoluble lipids, minimize the surface exposed by the lipids and have a null tension (Figure 4.8B), which prevents them from bursting like soap bubbles (Section 6.3). On the other hand, they can be stretched (Figure 4.8A) by creating an osmotic difference. They are then spherical like drops and bubbles.

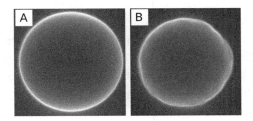

FIGURE 4.8 (A) Membrane of a tense vesicle. (B) Membrane relaxed under zero tension. These objects are a few tens of microns in radius. (Courtesy of O. Sandre.)

#### 4.2.1.1 Physical Origin

The region that separates two phases (e.g. liquid water and vapor) or two immiscible liquids (for example water and oil) is called the interface. A liquid interface (usually considered infinitely thin) is described by its surface area $A$, and associated with an energy, the surface tension. To minimize the surface energy of the system, $A$ must be minimal. The physical origin of the surface tension lies in the intermolecular forces involved in the cohesion of liquids (van der Waals, hydrogen bonds). Let us take a liquid–vapor interface: in the bulk liquid, the molecules benefit from attractive interactions with all their neighbors. At the surface, they lose half of the interactions (Figure 4.9).

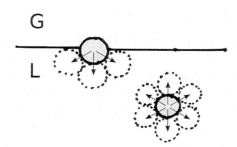

FIGURE 4.9 A molecule sitting at a surface loses half of its interactions with neighboring molecules.

Surface tension, or interfacial tension, $\gamma$ directly measures this energy loss per unit area. If $a$ is the molecular dimension, $a^2$ is the surface area per molecule, and $U$ is the cohesion energy, then:

$$\gamma \approx \frac{U}{2a^2}$$

Typical values for the surface tension span from $\sim$ 1 to $\sim$ 100's mJ/m$^2$ (Table 4.1).

TABLE 4.1    Values of $\gamma$ the Surface Tension for Different Liquids

| Liquids | Ethanol | Glycerol | Water | Water/Oil | Mercury |
|---|---|---|---|---|---|
| $\gamma$ (mJ m$^{-2}$) | 23 | 63 | 72 | 50 | 485 |

For a van der Waals liquid, with $U \sim k_B T$, $\gamma \sim 20$ mJ m$^{-2}$; for a metal like mercury, $U \sim 1$ $eV$, $\gamma \sim 500$ mJ m$^{-2}$.

Note that surface tension also varies with temperature. Generally, $\gamma$ decreases when $T$ increases ($d\gamma / dT < 0$). Because the cohesion energy decreases as thermal energy increases, the liquid becomes less dense. For example, the surface tension of water is 75.64 mN/m at 0.01°C, 72.75 mN/m at 20°C and 59.87 mN/m at 95°C.

### 4.2.1.2 Mechanical Definition of Surface Tension
Mechanical energy is required to produce emulsions of water in oil and therefore create interfaces.

The work $dW$ necessary to increase the surface $A$ by $dA$, which is proportional to the number of molecules brought to the surface (thus to $dA$), defines $\gamma$:

$$dW = \gamma\, dA$$

Dimensionally, $[\gamma] = EL^{-2} = FL^{-1}$. The unit is mJ m$^{-2}$ or mN m$^{-1}$. So $\gamma$ is the energy required to increase the area by one.

In surface thermodynamics, the work is related to variations of volume and surface: $dW = -pdV + \gamma dA$ and the variation of the free energy, $dF = -pdV + \gamma dA - SdT$ leads to the thermodynamic definition of the surface tension:

$$\gamma = \left.\frac{\partial F}{\partial A}\right|_{T,V}.$$

### 4.2.1.3 Capillary Forces

To illustrate surface tension, P.-G. de Gennes was used to starting with the paintbrush experiment. A dry paintbrush is shaggy, whereas a wet paintbrush looks like a thin and smooth bundle of hairs. De Gennes further invented the two-hair paint brush to describe the origin of capillary forces (Figure 4.10A). In a more simple geometry, let us take a rectangular frame and a moving rod dipped in soapy water. The shaded area represents the soap film in Figure 4.10B. To move the rod by $dx$, one has to perform the work $dW = F\,dx = 2\,\gamma\,l\,dx$, where the factor 2 comes from the fact that there are two water–air interfaces. This relation shows that $\gamma$ is also the force per unit length, which is exerted in the plane of the surface and directed towards the liquid.

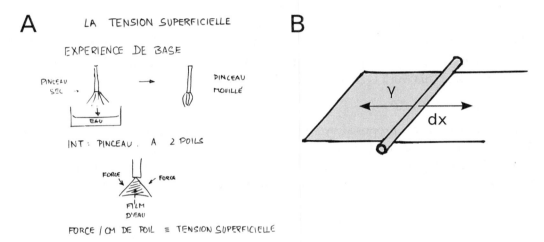

FIGURE 4.10   (A) De Gennes' introduction to surface tension (in French) based on the comparison between the dry and wet paintbrush and the two-hair paintbrush experiment. (B) Rectangular frame and mobile rod dipped in soapy water.

These forces are called capillary forces. They play remarkable roles. For instance, they allow you to hang toothbrushes on the wall; they make your hair stick together when you take a shower. Insects can also walk on the surface of water thanks to capillarity. But, if you add surfactant, the surface tension decreases, and they sink. This can be checked by carefully placing a steel needle on the surface of the water with a cigarette paper. Even when you remove the paper, despite its density it floats. When you add a drop of dishwashing liquid, you make it sink.

### 4.2.2 Laplace Formula (1805)

The surface tension is at the origin of the pressure which is exerted inside a drop or a bubble (Figure 4.11). It describes the pressure jump at curved interfaces.

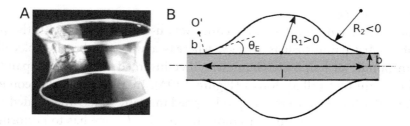

FIGURE 4.11 **(See color insert.)** (A) Soap film between two rings: zero curvature surface. (K. Mysels). (B) Unduloidal shape of a drop on a fiber.

### 4.2.2.1 Sphere

Let us consider a drop of oil in water (Figure 3.12A). To lower its surface energy, the drop takes the shape of a sphere of radius $R$. The work to displace the surface by a radial length $dR$ is written $dW = -p_1 dV_1 - p_2 dV_2 + \gamma dA$ with $dV_1 = -dV_2 = 4\pi R^2\,dR$ and $dA = 8\pi R\,dR$.

We write that this work is zero because the energy of the system at equilibrium is minimal. So $dW = 0$ gives $\Delta P = P_1 - P_2 = \dfrac{2\gamma}{R}$.

This law, which was proposed by Laplace, shows that the pressure increases when the size of the drops decreases. For this reason, in polydisperse emulsions, small drops disappear in favor of big ones. Along the same lines, if two bubbles of different size are connected by a straw, the small one ends up flowing into the big one (Figure 4.12B).

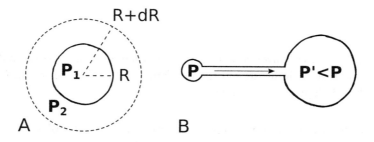

FIGURE 4.12   Laplace pressure. (A) Notations. (B) Illustration.

### 4.2.2.2 Generalized Surface

Let us take a pear-shaped object to define the curvature.

To determine the curvature $C$ of a surface at a point M (Figure 4.13), we draw the normal $\vec{N}$ at this point. The intersection of the surface with two perpendicular planes intersecting each other along $\vec{N}$ defines two curves of radius of curvature $R_1$ and $R_2$. If the center of the circle is inside the object, then $R > 0$, and outside, $R < 0$. Thus, for the pear of Figure 4.13, we have $R_1 < 0$ and $R_2 > 0$. The curvature is defined by:

$$C = \frac{1}{R_1} + \frac{1}{R_2}.$$

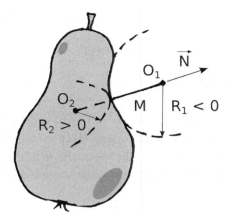

FIGURE 4.13  Radii of curvature of a pear: a needle is pushed into a point $M$ perpendicular to the surface. The pear is cut along two perpendicular planes intersecting along $\vec{N}$.

The Laplace pressure is written in a general way:

$$\Delta P = \gamma C = \gamma \left( \frac{1}{R_1} + \frac{1}{R_2} \right).$$

### 4.2.2.3 Surface with Zero Curvature

4.2.2.3.1 Meniscus on a Fiber   For thin fibers with a radius smaller than one millimeter, the shape of the meniscus is controlled by the surface energy. The profile is calculated by writing the conservation of the capillary forces (with the notations of Figure 4.14):

$$2\pi b \gamma = 2\pi z \gamma \cos\theta$$

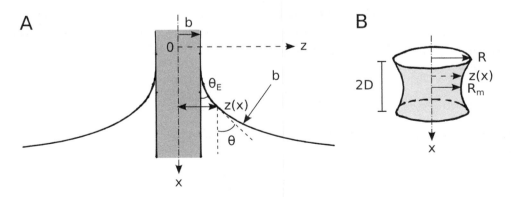

FIGURE 4.14   (A) Meniscus on a fiber. (B) Soap film between two rings separated by a distance $2D$ (drawing corresponding to Figure 4.11A).

By expressing $\cos\theta$ as a function of $\tan\theta$, we obtain the equation:

$$z\Big/\sqrt{1+(dz/dx)^2} = b$$

and the solution is: $z = b\cosh(x/b)$.

4.2.2.3.2 Soap Films    By dipping two circular rings of radius $R$ into a solution of soapy water and separating them gently to a distance of $2D$ (Figures 4.11A and 4.14B), a soap film is formed. As there is air communication between the inside and outside, the pressure is identical on both sides, and a surface with zero curvature is generated:

$$\Delta P = 0, \quad \left(\frac{1}{R_1} + \frac{1}{R_2}\right) = 0$$

The analytical expression of the profile $z(x)$ is the same as the one calculated for the fiber:

$$z = R_m\cosh(x/R_m)$$

where $R_m$ is the radius and the constriction is given by:

$$R = R_m\cosh(D/R_m)$$

This equation has two solutions for $R_m$, both corresponding to surfaces with zero curvature. The surface area is minimum for the first one and maximum for the second one. For a critical value of $R/D$, the two solutions become identical. For $R/D \le 1.5$, solutions no longer exist and the film bursts.

### 4.2.3 Capillary Adhesion

Two wet surfaces can stick strongly together. If the liquid wets the surface – contact angle $\theta_E$ less than $\pi/2$ (Section 4.4) – the Laplace pressure in the droplet between the two plates is negative. If $\theta_E = 0$, the Laplace pressure $\Delta P$ is given by

$$\Delta P = \gamma\left(\frac{1}{R} - \frac{2}{H}\right) \approx -\frac{2\gamma}{H}.$$

This pressure creates an attractive force:

$$F = \pi R^2\left(-\frac{2\gamma}{H}\right).$$

For $\gamma = 70$ mN m$^{-1}$, $R = 1$ cm, $H = 5$ μm, one finds $F \approx 10N$ which corresponds to a weight of 1 kg (Figure 4.15).

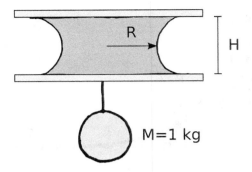

FIGURE 4.15   Force due to capillary adhesion.

On the other hand, if the liquid is non-wetting (contact angle greater than π/2), the surfaces repel each other.

References

1. H. Bouasse, *Capillarité. Phénomènes superficiels*. Impr. Paul Brodard, 1924.
2. P.-G. de Gennes, F. Brochard-Wyart, D. Quéré, *Gouttes, bulles, perles et ondes*. Belin, 2005.

## 4.3  CAPILLARITY AND GRAVITY

While small drops have spherical shapes, big ones flatten out. The action of both surface tension and gravity determines the shape of the drops. Capillarity is also at the origin of a singular behavior of liquids that seem to defy the law of gravity in the phenomenon of capillary rise.

### 4.3.1  Capillary Length

The capillary length, denoted by $\kappa^{-1}$, is the characteristic length above which gravity becomes dominant. It can be estimated by comparing the Laplace pressure $\gamma/\kappa^{-1}$ to the hydrostatic pressure $\rho g \kappa^{-1}$ at a depth $\kappa^{-1}$ in a liquid of density $\rho$ subjected to the gravity of the Earth $g$. The equality of these two pressures defines the capillary length:

$$\kappa^{-1} = \left(\frac{\gamma}{\rho g}\right)^{1/2}$$

For water, taking $\gamma = 70$ mN/m, $\rho = 10^3$ kg/m³ and $g = 9.8$ m²/s, we find $\kappa^{-1} \approx 3\,mm$.

Drops of radius $R \ll \kappa^{-1}$ are not affected by gravity and take the shape of a sphere in suspension or a spherical cap on a substrate. On the other hand, drops of radius $R \gg \kappa^{-1}$ are flattened by gravity and form puddles.

Due to capillary forces, a wetting liquid, such as water in a glass, forms a meniscus whose elevation from the mean level is $h \approx \kappa^{-1}$ (Figure 4.16). In a very narrow capillary of radius $R < \kappa^{-1}$, liquids climb spontaneously (Figure 4.17). This phenomenon is named capillary rise.

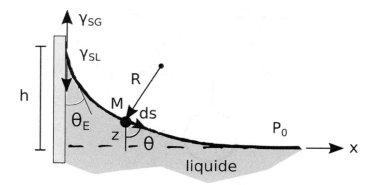

FIGURE 4.16  Sketch of a meniscus.

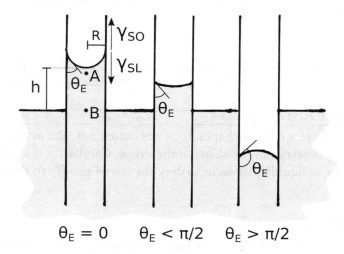

$$\theta_E = 0 \qquad \theta_E < \pi/2 \qquad \theta_E > \pi/2$$

FIGURE 4.17  Capillary rise for wetting ($\theta_E < \pi/2$) and non-wetting ($\theta_E > \pi/2$) liquids.

$\kappa^{-1}$ becomes large in zero-gravity conditions and at the interface between two liquids of similar density.

### 4.3.2  Capillary Rise – Jurin's Law

If a capillary is immersed in a wetting liquid (such as water), the liquid rises to a height $h$. If it is immersed in mercury, the liquid goes down. To determine $h$, the pressure at point $A$ just below the meniscus can be calculated (Figure 4.16):

1. *Using a hydrostatic reasoning*: At a depth $h$, the increase in hydrostatic pressure is $\rho gh$. At point $B$ located at the same altitude as the free liquid surface (which is at atmospheric pressure $P_0$), $P_B = P_A + \rho gh = P_0$.

2. *Using the Laplace formula*: If the radius of the capillary is such that $R \ll \kappa^{-1}$, the liquid–air interface is spherical: $P_A = P_0 - \left(2\gamma/R_m\right)$, where $R_m$ is the radius of curvature of the meniscus. If $\theta_E$ is the contact angle (Section 4.4 and Figure 4.16), the projection of $R_m$ on the horizontal axis is $R = R_m \cdot \cos\theta_E$.

By inserting the expression of $P_a$ (ii) into that of $P_b$ (i), we obtain $h = \dfrac{2\kappa^{-2}}{R}\cos\theta_E$. For a wetting liquid: $\theta_E < \pi/2, h>0$ (cases a and b, Figure 4.16).

For example, we obtain:

$h \sim 2$ m for $\theta_E = 0$, $\kappa^{-1} = 3$ mm, $R = 10$ μm.

For a non-wetting liquid: $\theta_E > \pi/2, h<0$. This is the case for mercury in a glass capillary (case c, Figure 4.16).

What carries the liquid? We may check that the weight of the elevated liquid is equal to the capillary force exerted on the line of contact: $\text{Weight} = \pi R^2 \rho g h = 2\pi R(\gamma_{SG} - \gamma_{SL})$, with $\gamma_{SG}$ and $\gamma_{SL}$ the solid:gas and solid:liquid interfacial tensions. Using the Young relation (Section 4.4), we obtain: $\text{Weight} = 2\pi R\gamma \cos\theta_E$.

This phenomenon is important in the field of imbibition of porous media such as rocks or powders. For instance, if we put natural cocoa in milk, we get lumps because it is not soluble. Chocolate powder is treated to dissolve spontaneously by capillary imbibition. Let us also mention the capillary rise of sap in plants (Section 8.5).

## 4.3.3 Menisci

When water is poured into a glass, water rises slightly along the edges and forms a meniscus. Thus any liquid contained in a large solid container has a horizontal free surface, except near the wall where it deforms over a distance $\kappa^{-1}$ (Figure 4.17). The liquid rises along the wall if it is wetting (water on glass) and goes down in the non-wetting case (mercury on glass).

The shape of the meniscus results from the balance between capillary forces and gravity. If $z$ is the height of the interface above the level of the bath, the pressure at a point $M$ below the surface leads to the equality between hydrostatic pressure and Laplace pressure:

$$P_M = P_0 - \rho g z = P_0 + \gamma/R$$

If $s$ is the curvilinear length, the radius of curvature is given by $ds = -Rd\theta$. Eliminating $ds$ related to $dz$ through $ds\cos\theta = dz$, we find after integrating with the boundary condition that $z=0$ and $\theta=\pi/2$ at infinity:

$$\frac{1}{2}\kappa^2 z^2 = 1 - \sin\theta$$

We find the height of the meniscus $h$ by writing that the contact angle to the wall is $\theta_E$:

$$\frac{1}{2}\kappa^2 h^2 = 1 - \sin\theta_E.$$

For $\theta_E = 0$, we find $h = \sqrt{2}\kappa^{-1}$.

The weight of the displaced liquid is carried by capillary forces: Weight $= \int \rho g z dx = = \gamma_{SG} - \gamma_{SL}$.

Note: The size of the menisci becomes important in zero gravity and at the interface between two iso-density liquids.

### 4.3.4 Shape of Drops

Here we study the shape of drops placed on a solid substrate or floating on a liquid substrate.

#### 4.3.4.1 Drop on a Solid Substrate

Figure 4.18 shows the typical shape of drops when volume increases.

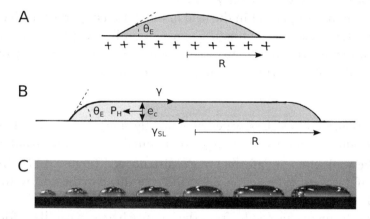

FIGURE 4.18 Shape of drops of increasing size. Figure 4.18 Shape of drops of increasing size. (A) $R < \kappa^{-1}$, spherical cap. (B) $R > \kappa^{-1}$, puddle or "pancake." (C) Illustration of the shape of drops of increasing size. (M. Fermigier).

For a small drop of radius $R < \kappa^{-1}$, the pressure in the drop is the Laplace pressure. As the pressure is uniform, the curvature is constant, and the shape of the drop is a spherical cap (Figure 4.18A).

We are now interested in the case of a larger drop, that is to say of radius $R > \kappa^{-1}$. The drop is flattened by gravitational forces and becomes a "pancake" (Figure 4.18B).

To derive the thickness of pancake-shaped drops, we write the equilibrium of forces (per unit of length) on a portion of the drop (Figure 4.18B, shaded area):

$$\gamma_{SO} + P_H = \gamma_{SL} + \gamma$$

with $P_H = \int_0^{e_c} \rho g z \cdot dz = \frac{1}{2} \rho g e_c^2$.

$P_H$ is the force due to the hydrostatic pressure integrated over the thickness $e_c$ of the liquid, and $\gamma_{SO}$, $\gamma_{SL}$, and $\gamma$ are the interfacial forces at the solid–air, solid–liquid, and liquid–air interfaces respectively. In setting $S = \gamma_{SO} - (\gamma_{SL} + \gamma)$ (Section 4.4), we find $\frac{1}{2} \rho g e_c^2 = -S$.

By using Young's relation: $\gamma_{SO} = \gamma_{SL} + \gamma \cos\theta_E$, which describes the balance of the capillary forces at the contact line, the thickness of the pancake can be written:

$$e_c = 2\kappa^{-1}\sin\frac{\theta_E}{2}.$$

Typically, $e_c = 6$ mm for a water drop on a non-stick pan ($\theta_E = \pi$). The thickness of this gravity puddle is also the critical thickness of dewetting of a water film deposited on a non wettable substrate, as shown in Figure 4.42.

We may also use an argument based on energy to derive $e_c$. The energy $F_g$ of the drop as a function of its surface area $A_g$ is: $F_g = F_0 - A_g S + \frac{1}{2}\rho g e^2 A_g$. By minimizing $F_g$ with the condition of constant volume (i.e. $A_g e = $ constant), we find $e = e_c$.

### 4.3.4.2 Floating Drop

A drop of liquid B is deposited at the surface of a liquid A. It can be done by pouring a spoon of oil onto the surface of water. As more oil is deposited, the lens flattens out. On a liquid substrate, the contact angle is not given by Young's relation but by Neumann's construction (Section 4.4). The contour of the drop is perfectly circular because a liquid surface is defect-free, smooth, and chemically homogeneous.

The thickness of the liquid lens can be calculated by the balance of forces. By definition, $S = \gamma_A - (\gamma_B + \gamma_{AB})$ is the spreading parameter of liquid B on liquid A.

As shown in Figure 4.19, hydrostatic pressures are equal in N and Q, which leads to:

$$\rho_B e = \rho_A e_1$$

Force balance on the hatched area is written: $P_1 + \gamma_B + \gamma_{AB} = P_2 + \gamma_A$, where $P_1 = \rho_A g e_1^2 / 2$ and $P_2 = \rho_B g e^2 / 2$ are the forces resulting from hydrostatic pressure.

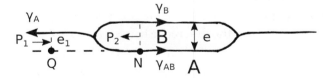

FIGURE 4.19   Floating drop.

We find: $-S = \frac{1}{2}\tilde{\rho}g e^2$, where $\tilde{\rho} = \rho_B\left(1 - \rho_B / \rho_A\right)$.

## 4.4 WETTING

Wetting [1] refers to the study of how liquid droplets deposited on a solid or a liquid substrate spread or do not spread. More generally, wetting deals with interfaces where three phases are in contact, most often liquid–solid–gas or liquid A–liquid B–gas. The substrate may be a smooth or rough surface or a porous medium.

Wetting is a phenomenon of everyday life that can be observed on rainy days: water pearls on a leaf, string of droplets on a spider's web, drops of water hanging on the windows are common and beautiful illustrations (see Figure 4.20).

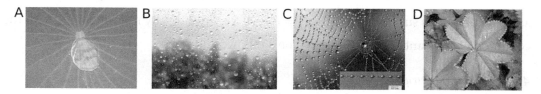

FIGURE 4.20 **(See color insert.)** Examples of wetting in everyday life. (A) Pearl of water on a lotus leaf (photo K. Guevorkian). (B) Drops of water anchored to a window. (Copyright Shutterstock.) (C) Spider web covered with dew (copyright Shutterstock) and zoom on a fiber with its rosary of water drops (F. Vollrath and Edmonds D.T., *Nature* (1989)). (D) Collar of drops on a "smart" leaf, which is hydrophobic to protect and hydrophilic to hydrate. (Photo FBW.)

But wetting is also of great industrial importance in many processes that require the spreading of a liquid on a solid. Whether it is a paint, a sunscreen, a lubricating film, it must be spread out in a stable film and avoid the formation of holes in the film, corresponding to the phenomenon of dewetting (Section 4.7).

### 4.4.1 Spreading Parameter $S$

When a drop of water is deposited on clean glass, it spreads out completely. On the other hand, the same drop deposited on a plastic sheet remains cohesive and does not spread out. The parameter that distinguishes both wetting regimes shown in Figure 4.21 is the spreading parameter $S$, which measures the difference between the surface energy of dry and wet substrates: $S = E_{dry} - E_{wet}$. Equivalently, $S = \gamma_{SO} - (\gamma_{LS} + \gamma)$, where $\gamma_{SO}$, $\gamma_{LS}$, and $\gamma$ are the surface tensions at the solid–air, solid–liquid, and liquid/air interfaces, respectively.

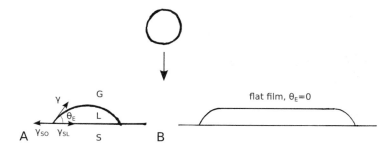

FIGURE 4.21 The spreading parameter $S = \gamma_{SO} - (\gamma_{SL} + \gamma)$ defines the two wetting regimes for a drop on a substrate. (A) Partial – Young (1805) [2]. (B) Total – Joanny, De Gennes (1984) [3].

These interfacial tensions between the three phases solid/liquid/gas (S/L/G) are the capillary forces per unit length acting on the contact line (also called the triple line).

### 4.4.2 Partial Wetting: $S < 0$

#### 4.4.2.1 Ideal Surfaces

Wetting is unfavorable so that a small drop will reduce contact with the surface, leading to a truncated sphere resting on the substrate at equilibrium, with a contact angle $\theta_E$ (Figure 4.22A).

This angle is given by the *Young relation* [2]: by projection of the capillary forces acting onto the contact line on a solid substrate, we obtain the Young–Dupré relation (1805)

$$\gamma \cos\theta_E = \gamma_{SO} - \gamma_{SL}. \tag{4.1}$$

This relation can also be found by writing that the work of the capillary forces when we move the line of contact by $dx$ is zero because the system is in equilibrium (($\gamma_{SO} - \gamma_{SL} - \gamma \cos\theta_E) \, dx = 0$).

The measurement of $\theta_E$ gives $S$: $S = \gamma(\cos\theta_E - 1)$

The particular case $\theta_E = \pi$ corresponds to no wetting. This case is encountered with a superhydrophobic substrate such as a glass slide coated with carbon black, a lotus leaf, or a duck feather (Section 8.1). The drop forms a pearl with a contact point with the surface only. Conversely, $\theta_E = 0$ corresponds to the wetting transition between the partial and total wetting situations.

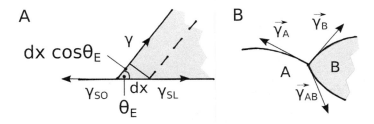

FIGURE 4.22 Triple line (A) S/L/G – Young relation and (B) $L_A$/$L_B$/G – Neumann's construction [4].

If the drop of liquid $B$ is deposited on a liquid substrate $A$, $S = \gamma_A - (\gamma_B + \gamma_{AB})$. The condition of mechanical equilibrium at the triple line (Figure 4.22B) is given by: $\vec{\gamma}_A + \vec{\gamma}_B + \vec{\gamma}_{AB} = \vec{0}$.

#### 4.4.2.2 Hysteresis of the Contact Angle

We often see that the drops remain anchored on a window. This is due to the hysteresis of the contact angle. Indeed, on a real (rough, dirty) surface we measure an advancing angle (at the front of the drop on an inclined surface or when pumping liquid into the drop) and a receding angle (at the back of the drop on an inclined plane or when removing liquid from the drop) (Figure 4.23). The force per unit length that holds the drop and avoids rolling is given by $\gamma(\cos\theta_r - \cos\theta_a)$. We find that on a smooth and chemically homogeneous model surface where $\theta_a = \theta_r = \theta_E$, this force is zero and the drop rolls.

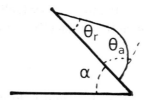

FIGURE 4.23 Hysteresis of contact and definition of advancing and receding angles.

The measure of contact angle hysteresis is a test of the cleanliness of a surface, which is used in the car industry before painting.

### 4.4.3 Complete Wetting: $S > 0$

If the parameter $S$ is positive, the fluid will spread over a large part of the surface in order to lower its surface energy. The equilibrium state is a film of nanoscopic thickness, called a van der Waals pancake [3], resulting from the balance between molecular and capillary forces (Section 2.3).

The energy $F_{film}$ per unit area of the solid covered by a film of thickness $e$ is (Figure 4.24): $F_{film}(e) = \gamma_{SL} + \gamma + P(e)$ where $P(e)$ is a corrective term for microscopic films, and $\gamma_{SL}$ and $\gamma$ are defined for semi-infinite media. $P(e)$ satisfies the following boundary conditions: $P(0) = -S$ and $P(\infty) = 0$. Furthermore, the disjoining pressure, which is the pressure acting on the liquid film due to the interactions with the solid substrate, is defined as $\pi(e) = -(dP/de)$.

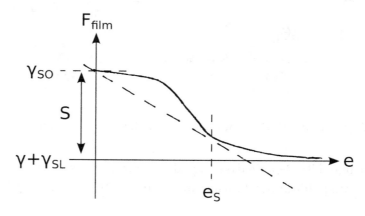

FIGURE 4.24 Energy $F_{film}(e)$ of a liquid film corresponding to the total wetting regime.

The thickness of the "van der Waals pancakes" is obtained by minimizing $F_{crêpe} = F_0 - SA + P(e)A$ with respect to the area $A$ at constant volume. We find the relation yielding $e_S$:

$$S = e_S \, \pi(e_S) + P(e_S). \tag{4.2}$$

The curve in Figure 4.24 displays $F_{film}(e)$ and shows the construction of the tangent that reveals $e_S$. For a fluid in which the dominant molecular interactions are van der Waals forces, such as oil,

$$P(e) = \frac{A_H}{12\pi e^2}$$

where $A_H$ is the Hamaker constant (Section 2.3). We then obtain:

$$e_S = a\left(\frac{3\gamma}{2S}\right)^{1/2}$$

where $a$ is the molecular size defined by $A/6\pi = \gamma a^2$. If $\gamma/S \approx 1$, $e_S \sim a$; if $\gamma/S \approx 100$, $e_S \sim 10a$ (thick pancake). If $S \to 0$, $e_S \to \infty$, this is the wetting transition.

For water, a wettable surface may also be referred to as *hydrophilic* and a nonwettable surface as *hydrophobic*.

## 4.4.4 From Wetting to Adhesion

Wetting can be extended to the description of the bonding or adhesion between two materials. Adhesion is the physicochemical phenomenon that occurs when two materials are brought into contact. Once the contact is established, the energy required to separate the materials is called the Dupré adhesion energy (Figure 4.25).

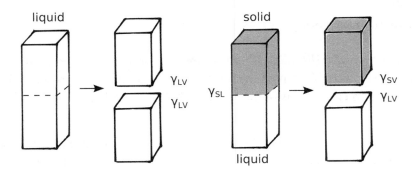

FIGURE 4.25 Diagram of different cases of cleavage. Left: liquid → liquid + liquid; right: solid/liquid → liquid + solid.

The adhesion work is the energy required to create two new surfaces from one interface.

If a homogeneous material (here, the liquid) is cleaved, the new surfaces are identical (Figure 4.25, left). The work of cleavage, called cohesion work, is $W_{cohesion} = 2\gamma_{LV}$.

If we cleave two different materials (Figure 4.25, right), $W_{adhesion} = \gamma_{SV} + \gamma_{LV} - \gamma_{SL}$, where $\gamma_{SV}$ and $\gamma_{LV}$ are the surface energies of the two new surfaces and $\gamma_{SL}$ is the interfacial tension.

$S$ can thus be related to the difference between adhesion and cohesion energy:

$$S = \gamma_{SL} - (\gamma_{LV} + \gamma_{SL}) = W_{adhesion} - 2\gamma_{LV} = W_{adhesion} - W_{cohesion}.$$

The main adhesion mechanisms to explain why one material sticks to another are as follows: (1) *structural adhesion*: adhesive materials fill voids or pores in surfaces and inter-lock the surfaces; (2) *chemical adhesion*: two materials form a new compound when assembled; (3) *long-distance forces*: two materials are kept in intimate contact by van der Waals forces; (4) *electrostatic adhesion*: for surfaces with opposite charge; and (5) *adhesion by interdiffusion*. For example, two polyethylene (PE) pipes are glued together by heating the junction: the PE chains interdigitate and heal the interface.

References

1. P.-G. de Gennes, *Rev. Mod. Phys.*, 57, 827 (1985).
2. T. Young, *Philos. Trans. R. Soc. London*, 95, 65 (1805).
3. J.F. Joanny, P.-G. de Gennes, *C. R. Acad. Sci., Paris Serie 2*, 299, 279 (1984).
4. F. Neumann, Vorlesungen über die Theorie der Capillaritt, (B.G. Teubner, Leipzig,1894).

## 4.5 PHYSICAL CHEMISTRY OF WETTING – ZISMAN CRITERION AND SURFACE TREATMENT

Physicochemistry of wetting aims to understand why a drop of water spreads on glass and not on plastic, and to find methods that allow us to make the drop spread on plastic and not on glass. In general, it is important to know how to predict and control the wettability of a substrate.

### 4.5.1 Wetting Criterion: Sign of the Spreading Parameter

There are two main categories of solids:

- *The "high energy" (HE) solids*: These are metals, covalent or ionic crystals, and glass. They are very wettable, which means that most liquids spread completely on them. Their cohesion energy $U$ is of the order of that of a covalent bond, i.e. $U \sim$ eV (or 40 $k_B T$);

- *The "low energy" solids (LE)*: These include plastics or molecular crystals. They are generally less wettable surfaces, i.e. liquids partially spread onto them. Their cohesion energy $U$ is of the order of that of thermal agitation, i.e. $U \sim k_B T$.

One might think that their wettability is due to the fact that the surface energy $\gamma_{SO}$ is very high for HE solids, low for LE solids. But we have seen (Section 4.4) that the ability to predict whether the solid will be partially or totally wet depends upon the sign of the spreading parameter $S$.

An energy balance corresponding to the "healing" process of the media allows us to connect $S$ to the polarizability of the liquid L and of the solid S (Figure 4.26).

1. Before healing two solids (Figure 4.26A), the initial surface energy is $2\gamma_{SO}$. After healing, the interfacial energy is zero. Starting from the initial state and considering that sticking them back together allows us to gain an energy $V_{ss}$, corresponding to the

solid/solid van der Waals interactions, and the chemical binding energy $U$ between the atoms of the solid, one obtains:

$0 = 2\gamma_{S0} - V_{SS} - U$, with $V_{SS} = k\alpha_S^2$ where $\alpha_S$ is the polarizability of the solid, $U$ the energy of a chemical bond, and $k$ a constant.

2. Similarly, when two liquid media are healed (Figure 4.26B), an initial energy is used to recover the liquid/liquid van der Waals interactions $V_{LL}$. The energy balance is:

$0 = 2\gamma - V_{LL}$, with $V_{LL} = k\alpha_L^2$ where $\alpha_L$ is the polarizability of the liquid.

3. Using the same reasoning, the healing process between a solid and a liquid (Figure 4.26C) gives:

$$\gamma_{SL} = \gamma_{S0} + \gamma - V_{SL}, \text{ with } V_{SL} = k\alpha_S\alpha_L.$$

Finally, we get:

$$S = \gamma_{S0} - (\gamma_{SL} + \gamma) = V_{SL} - V_{LL} = k\alpha_S(\alpha_S - \alpha_L)$$

We note that if $\alpha_S > \alpha_L$ then $S > 0$; hence the rule: *a liquid spreads completely onto a solid if it is less polarizable than the solid.*

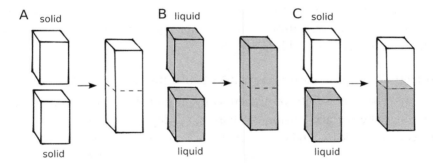

FIGURE 4.26 (A) Healing between two solids. (B) Healing between two liquid media. (C) Healing between a solid and a liquid.

## 4.5.2 Surface Treatment

The principle of surface treatment consists in changing the molecular nature of the surface of a solid to change its wettability properties.

### 4.5.2.1 How to Make a Wettable Surface Non-Wettable

A layer of less polarizable material than the liquid is applied to the surface.

Water wets clean glass ($S > 0$). If a hydrophobic molecular layer is grafted to the glass surface, the drop no longer spreads ($S < 0$). For instance, a trichlorosilane with a carbon chain R (e.g. $-(CH_2)_n$ or $-(CF_2)_m$) [1] that reacts with the Si–OH silanol groups on the glass surface can be used (Figure 4.27).

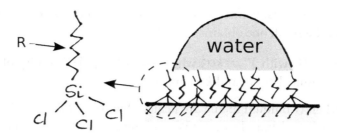

FIGURE 4.27 The water drop does not spread on the glass, because of the hydrophobic molecular layer. In the drawing on the right, the molecules are not represented to scale. Their size is a few nm while the drop can reach a few mm.

The application of this surface treatment allows windshields to remain dry, aircraft cabins to prevent frost build-up, frying pans to prevent food from sticking, and windows to become self-cleaning (Section 8.1).

### 4.5.2.2 How to Make a Non-Wettable Surface Wettable

A layer of material that is more polarizable than the liquid is applied, e.g. a metallic layer. This case corresponds to the pseudo-partial wetting regime (Figure 4.29C): a micro-drop forms a van der Waals pancake (thickness $e_s$). A large drop is surrounded by a nanoscopic wetting film (thickness $e_m$) that covers the surface.

One of the applications of this technique is in greenhouses, where condensation on non-wetting plastic deposits micro-droplets that scatter light. The plastic is made wet by sputtering a monoatomic layer of gold.

### 4.5.3 Surface Characterization – Zisman Critical Tension

To characterize the wettability of a chemically modified surface, drops of alkanes are deposited. Since the surface tension of alkanes, $\gamma_n$, depends on the number $n$ of groups ($CH_2$), the contact angle $\theta_E$ can be measured as a function of $\gamma$ (Figure 4.28). $\theta_E$ cancels for $\gamma = \gamma_C$, which is an intrinsic parameter characterizing the substrate and corresponding to $\alpha_L = \alpha_S$. $\gamma_C$ is called the Zisman critical tension [2].

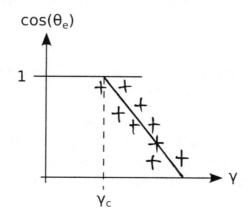

FIGURE 4.28 Zisman construction: Measure of $\cos \theta_E$ as a function of $\gamma$.

The determination of $\gamma_C$ allows the prediction of whether liquid of surface tension $\gamma$ will spread or not. It leads to the rule: If $\gamma < \gamma_C$, wetting is complete; If $\gamma > \gamma_C$, wetting is partial.

Examples: Naked glass, $\gamma_C \sim 150$ mN m$^{-1}$; Silanized glass, $\gamma_C \sim 20$ mN m$^{-1}$; Fluorintaed glass, $\gamma_C \sim 6$ mN m$^{-1}$. This high value for $\gamma_C$ on naked glass indicates that almost all liquids (except mercury, $\gamma = 485$ mN m$^{-1}$) spread on naked glass.

### 4.5.4 Wetting Criterion – Sign of the Hamaker Constant

Knowing the sign of $S$ is not sufficient to predict the wetting regime. It is also necessary to know the sign of the Hamaker constant: $A_H = A_{SL} - A_{LL}$ (Section 2.3), which controls the stability of the wetting films. Figure 4.29 shows the three main wetting regimes that are derived from the curves $F(e)$ characterizes the energy per unit area of a wetting film of thickness $e$. As seen in Section 4.4:

$$F(e) = \gamma_{SL} + \gamma + P(e),$$

with $P(0) = S$ and $P(e) = \dfrac{A_H}{12\pi e^2}$.

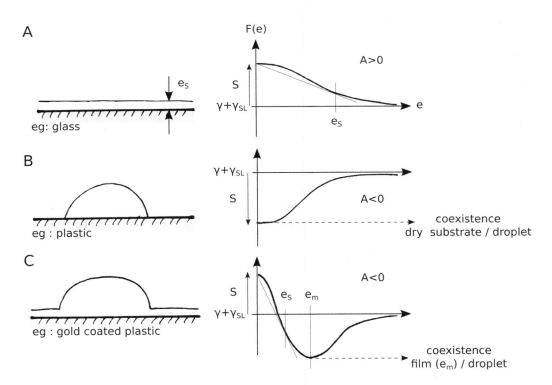

FIGURE 4.29 Different cases of wettability. (A) Total wetting. (B) Partial wetting. (C) Pseudo-partial wetting.

The three cases shown in Figure 4.29 are:

1. *Complete wetting*: $A_H > 0$ and $S > 0$, $\alpha_S > \alpha_L$, example: water on glass.

2. *Partial wetting*: $A_H < 0$ and $S < 0$, $\alpha_S < \alpha_L$, example: water on plastic.

3. *Pseudo-partial wetting*: $A_H < 0$ and $S > 0$, example: water on gold-coated plastic.

In case (C) a drop partially wets the surface but covers the rest of the substrate with a very thin film. For this reason, this situation has been called "pseudo-partial wetting."

References
1. J.B. Brzoska, N. Shahidzadeh, F. Rondelez, *Nature*, 360, 719–721 (1992).
2. W.A. Zisman, *Adv. Chem.*, 43, 1–51 (1964).

## 4.6 DYNAMICS OF WETTING

How does a drop spread? Why do raindrops that slide on windshields leave a trail of water? What is the speed of holes opening in a film that dewets? Why do we see drops moving on surfaces and even climbing slopes? Why do drops spread extremely quickly in partial wetting and take an infinitely long time in total wetting? These are all the questions that will be addressed here.

### 4.6.1 Capillary Velocity

An important parameter when dealing with the wetting of viscous liquids is the capillary velocity defined by:

$$V^* = \gamma / \eta$$

Where $\gamma$ is the liquid surface tension and $\eta$ its viscosity. If the liquid is not very viscous as for water ($\eta = 10^{-3}$ Pa s), $V^* = 72$ m·s$^{-1}$. Conversely, viscous liquids such as silicone oils are used to slow down the wetting dynamics for more accurate investigation. The capillary velocity $V^*$ can increase from 1 to $10^{-6}$ m·s$^{-1}$ when the length of the polymer chains, and thus its viscosity, increases (Section 7.6).

From $V^*$, we may estimate the characteristic time $T_e$ for the spreading of a drop in partial wetting, or for the fusion between two drops, using a dimensional argument:

$$V^* T_e = R_f$$

where $R_f$ is the final radius of the drop. For $R_f = 1$ mm, $T_e \sim 10$ microseconds for water, $\sim 1$ s for honey or a viscous oil, and $\sim 1$ h for ultra-viscous dough like *Silly Putty*.

The *capillary number* $C_a$ is a dimensionless number defined as the ratio of the speed of the triple line $V$ and $V^*$:

$$C_A = V / V^*$$

$C_a$ plays an important role in the dynamics of total wetting.

Another important parameter is the *Reynolds number* defined by:

$$Re = V L \rho / \eta$$

Where $\rho$ is the density of the liquid, $V$ is the speed, and $L$ is a characteristic size.

When $Re \ll 1$, the regime is viscous and controlled by $V^*$. For capillary phenomena, there is a transfer of mechanical energy (surface, gravitational) into viscous dissipation.

When $Re \gg 1$, the regime is inertial and the mechanical energy is transferred into kinetic energy. This is the case when soap bubbles burst or water dewets.

## 4.6.2 Dynamics: Partial Wetting

The dynamics of wetting is the study of the motion of contact lines.

### 4.6.2.1 Motion of the Contact Line

Young's relation characterizes the equilibrium contact angle $\theta_E$ when the triple line is at rest (Section 4.4). When the contact angle $\theta$ differs from $\theta_E$, the capillary forces are not compensated and the triple line moves at a speed $U$. The contact angle is called the dynamic contact angle $\theta_d$. The triple line advances if $\theta_d > \theta_E$, and the triple line recedes if $\theta_d < \theta_E$ (Figure 4.30).

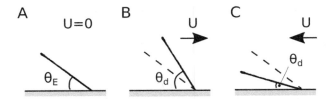

FIGURE 4.30   (A) Contact line at rest. (B) Advancing contact line. (C) Receding contact line.

The velocity $U$ of the line is the average of the velocity profile shown on the drawing.

The dynamics of wetting are given by the $U(\theta_d)$ law, which is obtained by writing that the driving force acting on the line is equal to the friction force.

*The driving force $F_M$ is the uncompensated capillary force:

$$F_M = \gamma_{SO} - \gamma_{LS} - \gamma \cos\theta_d = \gamma(\cos\theta_E - \cos\theta_d) \approx \frac{1}{2}\gamma(\theta_d^2 - \theta_E^2)$$

where we assume that the static and dynamic contact angles are small.

*The friction force is due to the friction exerted by the liquid flowing on the substrate (Figure 4.31).

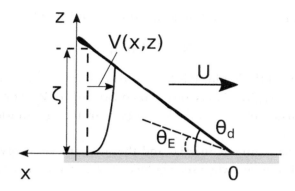

FIGURE 4.31   Flow in a moving liquid corner $\theta_d > \theta_E$.

Within the so-called lubrication limit, which assumes $\theta_d \ll 1$, the pressure in the moving liquid $P(x)$ is only dependent on $x$ and not on $z$. The flow in the wedge, characterized by the velocity $V(z)$, is a Poiseuille flow:

$$\eta \frac{\partial^2 V}{\partial z^2} = \frac{dP}{dx}$$

which meets the following boundary conditions:

$$V(0) = 0 \text{ and } \frac{\partial V}{\partial z}(\zeta) = 0$$

where $\zeta$ defines the profile of the liquid wedge. After integration, the solution that satisfies the boundary conditions is:

$$V = \frac{1}{2\eta} \frac{dP}{dx}\left(z^2 - 2z\zeta\right)$$

In the reference frame of the line moving at velocity $U$, the liquid is immobile,

$$\int_0^\zeta (V - U)\,dz = 0$$

hence:

$$U\zeta = \int_0^\zeta V(z)\,dz = \frac{-1}{3\eta}\frac{dP}{dx}\zeta^3$$

By eliminating $dP/dx$, the velocity profile is expressed as a function of $U$:

$$V = \frac{1}{2\eta}\frac{dP}{dx}\left(z^2 - 2z\zeta\right)$$

The viscous stress $\sigma_{xz}$ on the substrate is:

$$\eta\frac{\partial V}{\partial z}(z=0)=3\eta\frac{U}{\zeta}$$

The total force $F_v$ acting on the wedge, whose profile is $\zeta(x)=\theta_d x$, is given by:

$$F_v = \int_{x_{min}}^{x_{max}} 3\eta\frac{V}{\zeta(x)}dx = \int_{x_{min}}^{x_{max}} 3\eta\frac{U}{\theta_d}\frac{dx}{x} = 3\eta\frac{U}{\theta_d}\ln\left(\frac{x_{max}}{x_{min}}\right)$$

where $x_{max}$ is the drop size and $x_{min}$ is a microscopic length. In practice, $L_n=\ln(x_{max}/x_{min})$ is in the order of 20.

The equation $F_M=F_V$ can be expressed as:

$$\frac{1}{2}\gamma\left(\theta_d^2-\theta_E^2\right)=3\eta\frac{U}{\theta_d}L_n$$

And leads to the expression of $U(\theta_d)$ (Figure 4.32):

$$U=\frac{V^*}{6L_n}\theta_d\left(\theta_d^2-\theta_E^2\right) \tag{4.3}$$

This law is in agreement with experimental observations and allows the description of the dynamics of wetting, from drop spreading to dewetting.

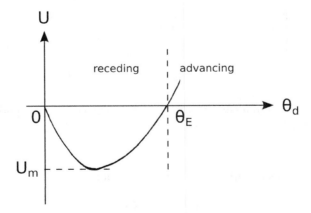

FIGURE 4.32 Velocity $U$ of the line as a function of $\theta_d$.

It is remarkable that the speed $U(\theta_d)$ cancels out for two values of $\theta_d$: for $\theta_d=0$, because the viscous dissipation in the wedge becomes infinite (the so-called "plumber's theorem") and for $\theta_d=\theta_E$ corresponding to the equilibrium of the contact line. There is a minimum speed $U_m$ between these two values. Below $U_m$, there is no solution for $\theta_d$: it is the regime of forced wetting. In practice, for water with $V^* = 72 \text{ m s}^{-1}$, $\theta_E = 0.1$ rad, and $L_n = 20$, one finds:

$$U_m = \frac{V^*}{9\sqrt{3}L_n}\,\theta_E^3 = 0.2\,\mu\text{m s}^{-1}$$

*Note:* we have assumed that the surface tension of the liquid is uniform. If the drop is deposited on a substrate with a thermal gradient, there is a surface tension gradient as we have seen that γ is a function of temperature. The liquid is drawn towards the high surface tensions. This liquid transport effect induced by a thermal gradient is called the *Marangoni effect* and can be found in many phenomena, such as the formation of wine tears along the walls of a glass.

### 4.6.2.2 Forced Wetting

When a plate is immersed vertically in a wetting liquid ($\theta_E < 90°$), the surface of the bath connects with the solid at an angle $\theta_E$ forming a meniscus (Figure 4.33A). When the plate is removed at low speed, i.e. $U < U_m$ (Figure 4.33B), the contact line is stable and the meniscus angle is the dynamic contact angle $\theta_d$. If the plate is removed at a speed $U > U_m$ (Figure 4.33C), the contact line is not stable and the plate carries a liquid film with it. This phenomenon is called *forced wetting*.

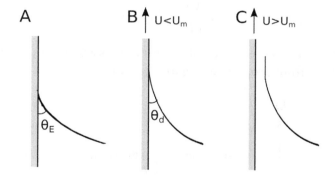

FIGURE 4.33  (A) Equilibrium meniscus when a plate is immersed in a liquid bath. (B) When the plate is removed at velocity $U < U_m$, the contact line is stable. (C) When $U > U_m$, the plate drives a liquid film.

The thickness $e(U)$ was calculated by Landau–Levich[1]:

$$e(U) = \kappa^{-1} \cdot Ca^{2/3}$$

and verified by D. Quéré [2] in a very simple and elegant experiment where he blows on a drop trapped in a Teflon tube. By measuring the decrease in the volume of the drop as a function of its speed, he could measure $U_m$ and the thickness of the film.

Let's go back to the raindrops that fall on the windshield. Driven at high speed by the moving car, they leave trails behind them: this is a simple demonstration of forced wetting.

An industrial application is the multi-layer treatment of optical lenses can be achieved by dipping the lenses in a liquid bath and by pulling them out at a velocity $U$ that will set the thickness of the layer.

### 4.6.2.3 Dynamics of Spreading of a Drop in Partial Wetting (S < 0)

Using the relationship $U(\theta_d) = dr/dt$ that describes the dynamics of the contact line, we derive the evolution of the radius $r(t)$ of a drop in partial wetting by assuming that the drop takes the shape of a spherical cap of radius $r(t)$ and height $h(t)$ (Figure 4.34). By writing the geometric relationship between $\theta$, $h$, and the volume of the drop $\Omega \approx hr^2$ and $\theta \approx h/r$, we find that the spreading dynamics of a drop in partial wetting is governed by:

$$r_E - r(t) = (r_E - r(0)) \cdot e^{-t/\tau}, \text{ with } \tau = 3L_n r_E / V * \theta_E^3 \text{ where } r_E \text{ is the equilibrium radius of the}$$

drop.

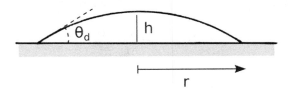

FIGURE 4.34   Spreading of a drop in partial wetting.

### 4.6.3 Dynamics: Complete Wetting

Compared to partial wetting, the profile of a drop in total wetting is completely different. A precursor film of a few nanometers thick extends at the front of the drop (Figure 4.35).

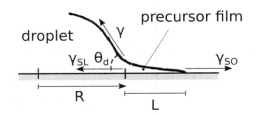

FIGURE 4.35   Profile of a liquid wedge advancing at speed $U$ in total wetting.

The precursor film results from the very high force acting on its end. The driving force is given by:

$$F_M = \gamma_{SO} - \gamma_{SL} \cos\theta_d = S + \gamma(1 - \cos\theta_d) \approx S + (1/2)\,\gamma\,\theta_d^2.$$

The total viscous force $F_V$ is the sum of the viscous contribution in the wedge of the liquid $F_{wedge}$ (similar to partial wetting) and the viscous contribution of the precursor film $F_{film}$: $F_V = F_{wedge} + F_{film}$. P.-G. De Gennes demonstrated that the contribution of $S$ in the driving force is compensated by the viscous resistance in the precursor film (i.e. by $F_{film}$) [3]. Thus, the balance between driving force and viscous dissipation leads to a balance between $F_{wedge}$ and $\gamma\theta_d^2/2$. This leads to Tanner's law:

$$\frac{3\eta U}{\theta_d} L_n = \frac{1}{2}\frac{\gamma}{\eta}\theta_d^2,$$

which gives:

$$U = \frac{V^*}{6L_n} \theta_d^3$$

Finally, the law describing the evolution of the radius $R(t)$ of the drop is derived from volume conservation:

$$\Omega \cong R^3 \theta_d$$

and from the relation $U(\theta_d)$ (Equation 4.4) which is written:

$$U = \frac{dR}{dt} = \frac{V^*}{6L_n} \frac{\Omega^3}{R^9}.$$

We obtain $R^{10} \cong V^* \Omega^3 t$ or, equivalently $\theta_d \sim t^{-3/10}$.

As $dR/dt \sim \theta_d^3$, spreading slows down as $\theta_d$ decreases. The dynamics in complete wetting are extremely slow compared to those in partial wetting.

It is important to keep in mind that this equation applies only when $R$ is smaller than the capillary length $\kappa^{-1}$. When $R > \kappa^{-1}$, drops are flattened by gravity.

### 4.6.4 *Star Wars* Application: The Force of the Meniscus

- The Dark Side: In cell biology, cells are seeded in a Petri dish and covered with culture medium that contains all the nutrients necessary for their growth and division. Nutrients are consumed and a red colored indicator turns yellow after a few days because the medium has become acidified. This tells us that we need to restore a fresh environment. With the advent of microfluidics (Section 2.5), many research groups have developed microfluidic systems for perfusion or recirculation of media in order to automate cell culture. However, one of the recurring practical issues in microfluidics is the formation of bubbles, either at the connection between the external pipes and the device or over time by gas dissolution (because the PDMS is permeable to gases). This is often detrimental, as cells are found to detach or die after bubbles have passed.

- The Bright Side: On the other hand, in the completely different field of integrated circuit production in the electronics industry, RAMs and microprocessors must be built in dust-free rooms, because particles larger than 100 nm = 0.1 μm may impair their proper functioning. In practice, however, it is often complicated to clean a surface of its smallest dirt. Gas jets (spray) or ultrasound are only effective to remove particles larger than 1 μm. This low limit is due to the fact that the dust adheres to surfaces (such as silicon wafers in the present case) by van der Waals forces. However, for a spherical particle on a plane, we have seen (Section 2.3) that $F_{\text{sphere-plan}} = A_H \left( R/6H^2 \right)$ (in the limit $H \approx R$), with $A_H$ the Hamaker constant, $R$ the radius of the dust particle assumed to be spherical, and $H$ the separation distance from the substrate. However, the forces required for particle removal often depend on the particle cross-section or volume, so they increase by a factor of 100 or 1,000 when the size of the particles decreases by only an order of magnitude. Serendipitous observations made by Leenars in 1988 [4] showed that the passage of bubbles allowed

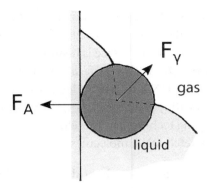

FIGURE 4.36 Drawing of a spherical particle adhered to a solid substrate (force $F_A$) and traversed by a bubble, i.e. a liquid–gas interface.

effective cleaning. The principle, schematized in Figure 4.36, is that the air–liquid interface creates a capillary force that (partially) opposes the adhesion force. As we know, the capillary force is $F_\gamma \approx \gamma R$, by forgetting the angular corrective factors. It therefore depends only on the size of the particle. Empirically, this was verified to be effective.

- The Combing of DNA: In 1994, A. and D. Bensimon used this capillary effect and the force exerted by the air/water meniscus to "comb" DNA [5]. DNA molecules in solution behave like a polymer coil (Section 10.1) and can be attached at one end to a microbead, a fiber, or to a glass slide (Figure 3.37A). The group from the École Normale Supérieure in Paris did this by treating glass slides with silanes (Section 4.5) carrying vinyl groups ($-CH=CH_2$) that have the peculiarity of interacting only with the ends of double-stranded DNA. When looking at a DNA solution squeezed between the silanized lamella and an untreated lamella, the DNA chains are therefore grafted as shown in Figure 4.37A. During the evaporation of the drop, it is observed that after the

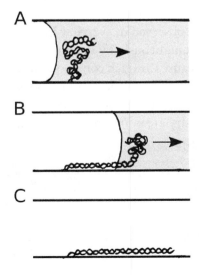

FIGURE 4.37 Drawing of a DNA molecule attached at one end to a surface and combed by the passage of a meniscus.

water–air interface has passed, the molecules are plated and stretched on the surface (Figure 4.37B), perpendicular to the meniscus of the receding drop. This means that the capillary force exerted by the interface is insufficient to break the DNA–glass bond but sufficient to stretch the DNA. This surface tension force can be estimated to about 400 pN by assimilating the DNA to a cylinder with a diameter $D = 2$ nm. However, as will be seen in Section 10.1, DNA undergoes a transition to an overstretched state around 70 pN. The trapping technique is therefore a simple method, based on capillarity, to immobilize and stretch DNA molecules (Figure 4.37C).

In molecular biology, fluorescence in situ hybridization (FISH) is a method that aims to visualize the spatial expression profile of different genes. Fluorescent oligonucleotides are injected into immobilized DNA to determine hybridization sites. Obviously, determination of this location is all the more accurate as the target DNA is stretched. This makes it possible, for example, to detect mutations or deletions in the DNA.

References

1. L. Landau, B. Levich, *Acta Physicochim. U.R.S.S.*, 17(42), 42–54 (1942).
2. De Ryck, D. Quéré, *J. Fluid Mech.*, 311, 219–237, 1996.
3. P.-G. de Gennes, *Rev. Mod. Phys.*, 57, 827 (1985).
4. A.F.M. Leenaars, in *Particles on Surfaces: Detection, Adhesion and Removal*, ed. K.L. Mittal, New York, NY: Plenum, 1988.
5. A. Bensimon, A. Simon, A. Chiffaudel, V. Croquette, F. Heslot, D. Bensimon, *Science*, 265, 2096–2098 (1994).

## 4.7 DEWETTING: WITHDRAWAL OF LIQUID FILMS

Dewetting, which refers to the spontaneous withdrawal of liquid films from a solid, liquid, or suspended substrate, is a phenomenon of everyday life that fascinated physicists and has many industrial applications.

### 4.7.1 Definition

Dewetting phenomena can be observed in everyday life (Figure 4.38). When we get out of the shower, our skin dries spontaneously by opening dry areas. Duck feathers are more effective than human skin because the duck comes out of the water completely dry. Before each flight, aircrafts are sprayed with a liquid that makes them super hydrophobic, to avoid the formation of an ice cocoon. In our kitchen, if we spread a film of oil on a non-adhesive frying pan, it is unstable, and dry areas are nucleated and grow.

A  B  C  D

FIGURE 4.38 Examples of dewetting: water on (A) the swimmer's skin, (B) the duck's feathers, (C) the aircraft cabin, (D) water or oil on a frying pan. (Copyright Shutterstock.)

If you are driving on a wet road, the water film between the tire and the road must be removed for adhesive contact to form and for the tire to have good grip, as sketched by

P.-G. de Gennes in Figure 4.39. Aquaplaning may occur when you brake suddenly: the contact gets wet, and you lose control of your car. By contrast, in the case of contact lenses, the intercalated film between the lens and the eye must be stable, otherwise the lens will stick unpleasantly to the eye.

FIGURE 4.39 **(See color insert.)** Aquaplaning sketched by P.-G. de Gennes. (FBW private collection.)

Dewetting also has industrial applications. Indeed, many industrial processes consist of spreading stable liquid films, and dewetting is undesired: this is the case with adhesive tapes, where the adhesive must coat the surface uniformly without leaving uncovered areas. Similarly, treatments for the farming industry are formulated from an aqueous base, but because leaves are very hydrophobic, active species must be added to increase the wettability. The formulation of rinsing products for dishwashers also takes into account this issue: it is important to avoid dishes and glasses being dry upon heating, leaving impurities stuck; if dewetting is favored prior to drying, impurities are carried away with the liquid. To avoid aquaplaning when driving on wet roads, the surface and texture of the tire must be treated so that the film between the tire and the road surface can dewet. Soles of sports shoes are textured and treated so that the athlete has a good grip on wet ground.

FIGURE 4.40 (A) Supported. (B) Intercalated between a hard solid and a soft rubber. (C) Suspended films.

Dewetting describes the rupture of a thin liquid film, whose viscosity can vary over several orders of magnitude. Three different geometries are usually distinguished (Figure 4.40):

1. Film deposited on a non-wetting substrate (either immiscible liquid or solid): This is the reverse process of spreading. When $S > 0$ (complete wetting regime), the liquid tends to cover the surface so that a film is always stable. When $S < 0$, the liquid tends to withdraw from the surface.

2. Film intercalated between a hard and soft solid, which controls adhesion.

3. Suspended film, which controls the bursting of bubbles and soap films (Section 6.3).

We will describe here the most common case of supported films.

### 4.7.2 Dewetting of Supported Films: Film Stability

We will first determine the critical thickness $e_c$ below which a film is metastable or unstable. We will then describe the dynamics of dewetting by nucleation and growth of dry areas.

We consider a liquid film of thickness $e$ deposited on a solid substrate with a spreading parameter $S < 0$ (Figure 4.41). $F(e)$ is the free energy of film per unit area.

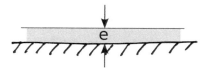

FIGURE 4.41  Liquid film of thickness $e$ on a solid substrate.

$F(e)$ is the sum of gravity energy $(= \rho g e^2 / 2)$ and interfacial energies $\gamma$ and $\gamma_{SL}$. In addition, in the absence of any film (i.e. for $e = 0$), we have $F(0) = \gamma_{SO}$.

$$F_{\text{film}}(e) = \gamma_{SL} + \gamma + \frac{1}{2}\rho g e^2$$

$$F_{\text{film}}(e = 0) = \gamma_{SO}$$

This function $F(e)$ (Figure 4.42) is not continuous close to zero: it has a discontinuity of magnitude $S$. This energy gain is the reason for dewetting.

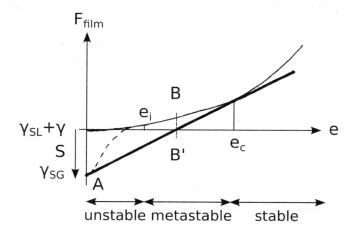

FIGURE 4.42  Energy of the liquid film $F_{\text{film}}$ as a function of thickness $e$.

The expression of $F_{\text{film}}(e)$ is valid for mesoscopic thicknesses, because $F_{\text{film}}(e \to 0) \neq F_{\text{film}}(0)$. To ensure continuity for $e \to 0$, long-range molecular forces must be included (Section 2.3).

Drawing the tangent to the curve from point A, we find an intersection for $e = e_c$. This so-called Maxwell's construction shows that there is a coexistence between the dry solid and a liquid film of thickness $e = e_c$, which corresponds to the thickness of a pancake deposited on the solid. For $e > e_c$, the film is stable. For $e < e_c$, the film is metastable. It lowers its energy (B→B') by dewetting through nucleation and growth of dry areas (Figure 4.43B). Thus a puddle of oil on a non-adhesive frying pan has a thickness $e_c$. If it is crushed with a spatula, holes/dry areas are formed. For thicknesses less than $e_i$ corresponding to the inflection point, of the order of 10 nm, the nanoscopic film is unstable and spontaneously breaks into numerous droplets. This regime, called spinodal decomposition, is shown in Figure 4.43A.

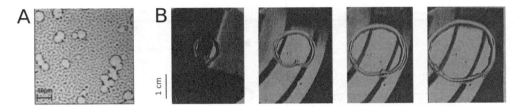

FIGURE 4.43  Dewetting (A) by spinodal decomposition, (B) by nucleation and growth of a dry area [1].

### 4.7.3 Dynamics of Dewetting

Depending on the viscosity of the liquids, which can range from 1 cP for water to $2 \times 10^9$ cP for ultraviscous pastes, three different dynamic regimes, namely inertial, viscous, and viscoelastic, can be observed. We will describe here the viscous regime, which is the most frequent and important in practice, and the inertial regime. The case of ultraviscous fluids will be discussed below (see Section 4.7.4).

#### 4.7.3.1 Viscous Regime

We will describe the dynamics of withdrawal of a film supported on a solid substrate. Figure 4.43B shows the nucleation and opening of a dry zone in a film of PDMS, a viscous silicone oil used to slow the phenomenon. The film was deposited by spin coating of a drop of PDMS on a silanized, non-wetting substrate (Section 4.5). The film is anchored to a wetting ring obtained by degradation of the surface treatment over a crown-shaped area by UV irradiation. At time $t = 0$, a hole is nucleated, and its opening dynamics are monitored by measuring the radius $R$ as a function of time. We see that the hole is surrounded by a liquid rim that collects the liquid from the dry area. The motion of the rim is driven by the force per unit length $\gamma + \gamma_{SL} - \gamma_{S0} - \frac{1}{2}\rho g e^2 \approx -S$ (within the limit of $e \ll e_c$).

To describe the opening of the hole, we use the law of movement of the contact lines $A$ and $B$ (Section 4.6), which border the rim (Figure 4.44A). It is assumed that pressure balance is very fast in the thicker parts of the rim and that its shape is a spherical cap. The contact angles in $A$ and $B$ are identical:

Line A: $V_A = V^* \theta_d \left( \theta_d^2 - \theta_E^2 \right)$

Line B: $V_B = V^* \theta_d^3$

By assuming $V_A = V_B = dR/dt$, we find $\theta_d = \theta_e / \sqrt{2}$ and $dR/dt = V^* \left( \theta_e^3 / 2\sqrt{2} \right)$.

Thus, the velocity of dewetting is constant. It is proportional to $\theta_e^3$, and varies as the inverse of viscosity. These two findings were experimentally verified by C. Redon using alkanes to vary the contact angle and silicone oils to vary the viscosity over several decades (Figure 4.44C) [1].

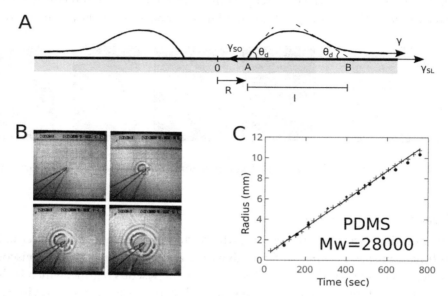

FIGURE 4.44  Dynamics of dewetting in viscous and inertial regimes. (A) Shape of the rim and notations. (B) Opening kinetics. (Courtesy of Claude Redon.) (C) Inertial dewetting of water on a hydrophobic surface ($t = 10$ ms between each image). (Courtesy of Axel Buguin.)

These laws on dewetting, which have been of practical importance, have been extended to very different fields. Examples also include cell adhesion, where the liquid film intercalated between the cell and the substrate must be removed to ensure proper cell adhesion, and the dynamics of macro-openings that are observed in epithelial cells infected with Staphylococcus aureus [2].

### 4.7.3.2 Inertial Regime

As viscosity decreases, dewetting velocity increases, and there is a transition towards inertial dewetting (Figure 4.44B). The dimensionless number that separates these two regimes is the Reynolds number $\mathrm{Re} = \rho V \ell / \eta$, where $\ell$ is the width of the rim.

In the inertial regime, the fundamental equation of mechanics applied to the rim is written: $d(MV)/dt = -S$. As the rim collects the liquid from the dry area, the mass is $M = \rho Re$. The solution of the fundamental equation is $V = \text{constant}$, $dM/dt = \rho Ve$, hence:

$$V = \sqrt{-S/\rho e}.$$

The burst speed in inertial regime is constant and varies as $e^{-1/2}$. This law, which was verified by A. Buguin [3] is identical to the one that describes the bursting of a soap film, if one replaces $-S$ with $2\gamma$, the surface tension of the film.

## 4.7.4 Application: Life and Death of Viscous Bubbles

In 1991, Pierre-Gilles de Gennes was rewarded with the Nobel Prize for his works on soft matter. The chemical and pharmaceutical company Rhône-Poulenc handed out plastic boxes labeled: "Do you know soft matter?" These boxes contained a magic dough that retained all its properties until now, as shown in the pictures in Figures 4.45 and 4.46! It is a pasty liquid. But if you shape it into a small sphere, you get an elastic ball that bounces back. If you let it rest, it spreads out like a liquid.

FIGURE 4.45    Spreading of a viscoelastic ball: it bounces when thrown, it spreads out when placed: (A) $t = 0$; (B) $t = 1$ h; (C) $t = 24$ h. (Photo FBW.)

When this dough is spread into a thin film, nucleated holes are visible and grow (Figure 4.46). Georges Debregeas, who was beginning his PhD on another subject at the time, was asked to record the opening of the holes. It turns out that this exciting side-project kept him busy for three years.

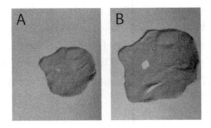

FIGURE 4.46    Spreading of the dough into a thin film and visualization of a hole that opens (A) and grows (B). (Photo FBW.)

### 4.7.4.1 Viscous Liquid Film: From Molten Polymer to Liquid Glass

A film of pure water is too unstable to be kept intact for more than one millisecond. In order to stabilize it, traces of surfactant must be added to reduce the surface tension. On the other hand, a film of liquid $10^8$ times more viscous is stable, because it flows so slowly that it takes a very long time to relax to an equilibrium state. This is why viscous foams can be made without surfactants. In the furnaces of the Saint-Gobain glass company, giant bubbles of molten glass can be formed and generate foams that hinder the proper functioning of processes. But there also exists an industrial process that takes advantage of this property and produces foams of expanded polymer by injecting gas into the polymer melt without adding surfactants. Finally, we can mention the bubbles that splash you when you heat up mashed potatoes or the viscous bubbles of lava or mud in a solfatar (Figure 4.47).

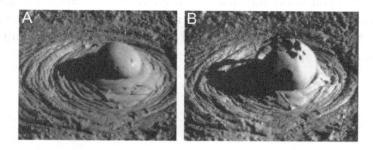

FIGURE 4.47   Formation (A) and bursting (B) of a viscous bubble in a solfatar in the United States, Yellowstone National Park (Katia and Maurice Kraft, vulcanologists).

While bursting of soap films has fascinated physicists for more than one hundred years, nobody had investigated bursting pasty liquids. Here, inertia is negligible, and film rupture is controlled by the viscoelastic properties of the dough.

To make thin films that are smooth at an atomic scale, the use of a rolling pin to crush pastry is inadequate. Instead, we dip a ring of radius $R$ in a polymer solution. Initially, the liquid film that is drawn is not very viscous, but as the solvent evaporates, a suspended film of molten polymer (of thickness $e$, typically 50 μm) can be formed. With a needle, you can puncture the film and record the growth of the hole. We observe that the film remains flat and that no rim is accumulated around the hole. The liquid is evenly distributed in the thickening film. Experimentally, we find that the growth law is exponential and that the opening velocity is proportional to 1) $V^* = \gamma/\eta$, where $\gamma$ is the surface tension and $\eta$ is the viscosity of the molten polymer ($\approx 10^8$ times that of water for the magic paste in Figures 4.45 and 4.46) and thus $V^*$ is extremely low, in the order of 10 μms$^{-1}$, and to 2) $R/e$ which is very high.

We describe the bursting laws here by writing that the surface energy gain is transformed into viscous dissipation. The main difference between a soap film and a suspended film is the nature of the flows: in the soap film, because of the surfactants on the surface, there is a Poiseuille flow, and the velocity cancels out at the edges of the film: viscous dissipation is huge for micrometric films. On the other hand, viscous films are bare, and the speed is the same throughout the whole thickness of the film: the flow becomes a so-called plug flow. Dissipation is then very low and very thin viscous films can burst as fast as soap films, whereas they are $10^8$ times more viscous!

The transfer of surface energy into viscous dissipation for a hole of size $R$ in a film of thickness $e$ which opens at velocity $dR/dt$ is written, by ignoring the numerical coefficients:

$$\eta \left( \frac{1}{R} \frac{dR}{dt} \right)^2 R^2 e \approx R\gamma \frac{dR}{dt} \tag{4.5}$$

The velocity gradient is here $\dfrac{1}{R}\dfrac{dR}{dt}$ instead of $\dfrac{1}{e}\dfrac{dR}{dt}$ as in a soap film, which leads to a growth that can be very fast if the film is thin.

From Equation 4.5, we derive that the growth law is exponential: $R = R_0 e^{t/\tau}$, with $\tau = \dfrac{\eta e}{2\gamma}$, and the opening velocity is proportional to $V^* \dfrac{R}{e}$. If $V^*$ is extremely small, $R/e$ can become very large. If we take $R = 1$ cm, and $e = 1$ micron, we find opening speeds of the order of mm/s.

### 4.7.4.2 Life and Death of Viscous Bubbles

After studying the bursting of suspended films, we investigated the case of viscous bubbles. By injecting a large air bubble into molten polymer, a drop rises and makes a beautiful bubble on the surface, with the shape of a half-sphere because the contact angle is 90°. This bubble becomes thinner due to drainage, and its thickness can be monitored by interferometric measurements. The bubble is then pierced with a needle, and its bursting is recorded, which can be so fast that an ultra-fast camera must be used [4]. The bursting of the hole in short times follows the exponential law demonstrated for suspended films; then when the hole has a size comparable to that of the bubble, its opening slows down, and we observe that it collapses via an instability like a parachute.

### 4.7.4.3 Exception: Transient Pores in Vesicles and Nuclear Membranes

When a vesicle or cell is swollen, pores are nucleated and grow because the membrane is stretched. Their openings are described by the model for the bursting of viscous films. But the vesicle or cell does not die: the pores close spontaneously (Figure 4.48). Indeed, as soon as a pore opens, the tension of the membrane is released by two mechanisms: the opening of the pore and the leakage of the inner liquid subjected to Laplace pressure. If the vesicle is immersed in a highly viscous liquid, macroscopic pores open and close slowly because the leakage of the inner liquid is slowed down. In water, the leakage is so rapid that the pores close before reaching an observable size. E. Karatekin and O. Sandre investigated the role of surfactants in the mechanical properties of the membrane and in the line energy associated with holes in the membrane [5]. Line energy is the two-dimensional equivalent of three-dimensional surface energy. Some surfactants are able to reduce this line energy by a factor of 100, making the membrane permeable. Several families of surfactants were tested according to their hydrophilic/hydrophobic balance and their affinity with being inserted into the membrane. High membrane permeability, as induced by tension or favored by the insertion of drugs, may have applications in drug transfer or allow us to gain insight into some fundamental biological processes (e.g. endocytosis).

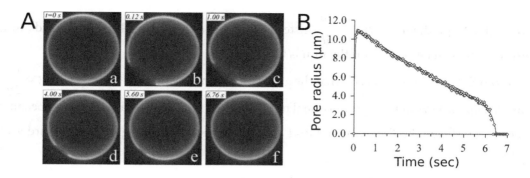

FIGURE 4.48 Opening and closing of membrane pores. (A) Sequence of video-photographs a–f. (B) Closing curve of a pore of size $r$ as a function of time. (Adapted from [3]. Courtesy of O. Sandre.)

In addition, living cells that migrate into the extracellular matrix are strongly deformed, and their nuclei are stretched. Transient pores appear in the nuclear membrane, which can lead to the loss of genetic material and cell death.

References

1. C. Redon, F. Brochard-Wyart, F. Rondelez, *Phys. Rev. Lett.*, 66(6), 715–718 (1991).
2. E. Lemichez et al., *Biol. Cell*, 105(3), 109–117 (2013).
3. A. Buguin, L. Vovelle, F. Brochard-Wyart, *Phys. Rev. Lett.*, 83, 1183 (1999).
4. G. Debrégeas, P.-G. de Gennes, F. Brochard-Wyart, *Science*, 279, 1704–1707 (1998).
5. E. Karatekin, O. Sandre, H. Guituni, N. Borghi, P.-H. Puech, F. Brochard-Wyart, *Biophys. J.*, 84, 1734 (2003).

# CHAPTER 5

# Liquid Crystals

## 5.1 GENERALITIES ON LIQUID CRYSTALS

### 5.1.1 Discovery

The Austrian botanist Friedrich Reinitzer, who studied the function of cholesterol in plants, discovered in 1888 that a cholesterol benzoate crystal, when heated, does not melt like an ordinary crystal but turns at 145°C into a milky liquid, which becomes perfectly clear at 178.5°C. Convinced that he had discovered a new state of matter, he sent it to Otto Lehman, who became famous for using the polarized light microscope in crystallography. Lehman had built a heating plate that allowed him to visualize the evolution of a crystal when it was heated, and he interpreted this milky state as an ordinary crystalline phase with a highly mobile three-dimensional lattice. He then extended Reinitzer's discovery to a large number of natural and artificial substances and, to recall these apparently contradictory characteristics, gave them the name "Liquid Crystal" in 1904. Reinitzer also showed that these liquid crystals have the property of rotating the polarization direction of light (Figure 5.1).

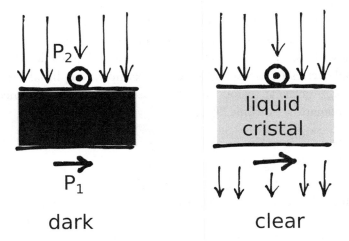

FIGURE 5.1  Between cross-polarizers, a liquid sample appears black and a liquid crystal bright.

At that time, the perplexity of the physicists, who refused to believe in a new state of matter, lasted until 1922, when Georges Friedel proposed the first classification of mesophases or intermediate states of matter between the liquid state and the crystalline state. He distinguished, in particular, between the nematic (or cholesteric) state and the smectic state. Later theoretical and experimental developments have been influenced by F.C. Frank (1950) and the Orsay Liquid Crystal group under the leadership of Pierre-Gilles de Gennes. For a complete review of the physics of liquid crystals see reference 1.

Liquid crystal phases are often obtained with elongated molecules consisting of a central rigid stick-shaped part that promotes the order of orientation and flexible ends preventing crystallization and promoting the liquid state (Figure 5.2).

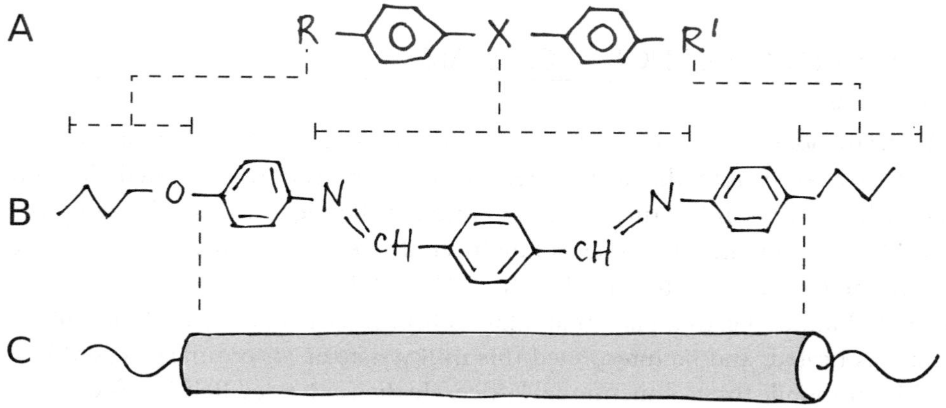

FIGURE 5.2   (A) Generic chemical structure of liquid crystals. (B) Chemical formula of the TBBA molecule. (C) Schematic representation: the central part is rigid, and the chains at both ends are flexible.

The long axis of the molecules is oriented along a mean direction, which can be fixed by the walls or an external force (e.g. magnetic or electric field). The result is an anisotropy of the physical properties that is comparable to that of crystals (e.g. optically active medium), and yet some mesophases (e.g. nematic) flow like ordinary liquids do. The same substance may have a whole series of mesophases. As an example, terephthal-bis 4n butylaniline (TBBA (Figure 5.2)) gives, at fusion, the sequence of phases: solid > smectic B > smectic C > smectic A > nematic > isotropic liquid.

## 5.1.2 Nematics

In the nematic phase, the long axis of the molecules possesses a privileged mean direction, characterized by a unit vector $\vec{n}$ (Figure 5.3), but the gravity centers of the molecules are randomly distributed, and a nematic flow like a liquid. The system is invariant by rotation around $n$ (uniaxial medium) and the states $n$ and $-n$ are identical. The system is uniaxial, and the order parameter introduced in Section 3.2 is a tensor:

$$S_{ij} = \langle n_i n_j \rangle - \frac{1}{3}\delta_{ij}$$

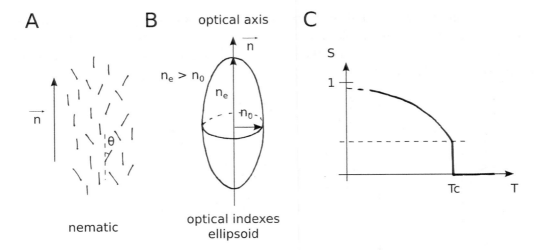

**A**

$\vec{n}$

nematic

**B**    optical axis

$\vec{n}$

$n_e > n_0$

$n_e$

$n_0$

$\theta$

optical indexes
ellipsoid

**C**

S

1

Tc        T

FIGURE 5.3   (A) The long axis of molecules is oriented on average according to the guiding vector $n$. (B) Optical birefringence properties of liquid crystals; $n_0$ and $n_e$ correspond to the ordinary and extraordinary refractive indices. (C) Typical shape of the nematic order parameter $S(T)$ which vanishes at the N→I transition temperature, $T_c$. $S(T)$ is discontinuous at $T_c$, which corresponds to a first order phase transition.

$S(T)$ measures the alignment rate along the director $n$ and is the order parameter of the nematic/isotropic transition $S(T) = \frac{1}{2}\langle 3\cos^2\theta - 1\rangle$. A perfect ordered state corresponds to $\theta = 0$, $S = 1$, and a disordered liquid to $\langle\cos^2\theta\rangle = 1/3$, $S = 0$.

The widely experimentally used nematics are PAA and MBBA, which are represented in Table 5.1.

TABLE 5.1   Chemical Structures of Two Nematics

| **Paraazoxyanisol (PAA)** <br> **nematics from 116 to 135°C** | **Methoxybenzylidene butyl anilin (MBBA)** <br> **nematics from 20 to 47°C ($T_{room}$)** |
|---|---|
| $CH_3 - O -\!\!\bigcirc\!\!- N=N -\!\!\bigcirc\!\!- O - CH_3$ <br> $\qquad\qquad\qquad\;\downarrow$ <br> $\qquad\qquad\qquad\;O$ | $CH_3 - O -\!\!\bigcirc\!\!- CH=N -\!\!\bigcirc\!\!\sim\!\!\sim$ |

### 5.1.3 Cholesterics

Cholesterics are a subclass of nematics. When the molecules are chiral, the director $\vec{n}$ rotates regularly around a privileged axis (Figure 5.4). This is the case for cholesterol derivatives, such as cholesterol benzoate studied by Reinitzer. The helical structure is characterized by the pitch of the helix "$p$," which varies typically from 0.1 to 10 μm, and by the direction of the right or left helix depending on whether the director rotates about the cholesteric axis counter clockwise or clockwise (right helix in Figure 4.4). A remarkable property of cholesterics is the selective reflection of light at a wavelength $\lambda_0 = np \cdot \cos\alpha$, where $n = (1/2)(n_0 + n_e)$ is the mean optical index and $\alpha$ the observation angle. If $\lambda$ differs from $\lambda_0$, the light is transmitted. Since the pitch of the helix $p$ is very sensitive to temperature, the cholesteric changes color and can be used as a temperature indicator. It is also this property that is at the origin of the iridescent colors of beetles whose shell has a fixed cholesteric order (Section 9.4).

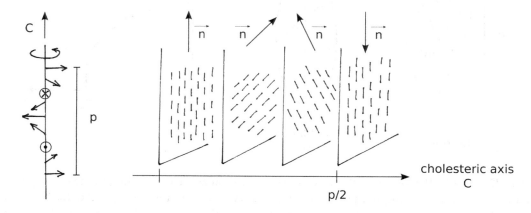

FIGURE 5.4 Helicoidal structure of cholesterics. Schematic representation of the director vector's rotation around the cholesteric axis. When a 180° rotation of $n$ is performed, a distance $p/2$ is covered along the cholesteric axis.

### 5.1.4 Smectics

The name of the smectic phases, given by G. Friedel, comes from the Greek *smectos*, soap.

In smectics, in addition to the direction of preferential alignment of molecules, there is an order of position of the centers of gravity: molecules are organized into layers (like soaps). A wide variety of smectic phases can be distinguished according to the degree of order in a layer, the correlations between layers, and the orientation of the alignment direction in relation to the normal vector at the layers' main direction.

The two most common smectic phases are shown in Figure 5.5. In a smectic A, each layer is a two-dimensional liquid, and the molecules align perpendicularly to the layers (uniaxial medium). In a smectic C, the molecules are inclined with respect to the normal at the layers, and the medium is biaxial (normal at the layers and molecular direction). We will not talk here about more exotic smectic phases such as the B, E, and F ones for the sake of simplicity.

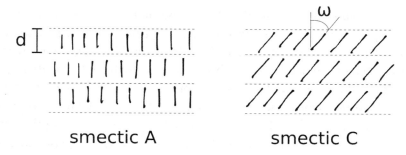

FIGURE 5.5 Most common smectic phases.

### 5.1.5 Thermotropic and Lyotropic Liquid Crystals

Thermotropic liquid crystals are observed by varying the temperature. But there exist also lyotropic liquid crystals that appear in solution (Figure 5.6). Water–lipid systems, when the water concentration varies, present a wide variety of phases, including lamellar phases isomorphic to smectic phases. Elongated rod molecules, semi-rigid polymers, carbon nanotubes, and some viruses also give rise to lyotropic nematic phases.

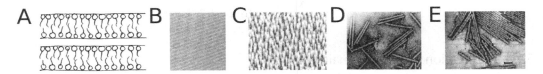

FIGURE 5.6 Different liquid crystal phases: (A) lamellar phase e.g. water/soap; (B) lyotropic nematic phase: DNA; nematic phase: (C) carbon nanotubes; (D) nematic phase: tobacco mosaic virus; (E) smectic phase: tobacco mosaic virus.

## 5.2 NEMATICS

Nematics are characterized by a unit vector $\vec{n}$ that describes the molecular orientation and by the alignment rate $S(T)$ of the molecules along this direction. Since the molecules are not polarized, $\vec{n} = -\vec{n}$, and the director is noted $n$.

### 5.2.1 Elasticity of Nematics: Frank–Oseen Theory [1, 2, 3]

When the director vector $\vec{n}$ varies spatially, it appears an energy of deformation $F$, which can be expressed in terms of the unit vector $\boldsymbol{n(r)}$. From symmetry arguments, the density of free energy can be expressed as a function of the three basic deformations that are splay, torsion, and flexion to which are associated three Franck's elastic constants $K_1$, $K_2$, $K_3$ (Figure 5.7).

$$F\big|_{m^3} = \frac{1}{2}K_1\left(div\,\vec{n}\right)^2 + K_2\left(\vec{n}.rot\,\vec{n}\right)^2 + K_3\left(\vec{n}\wedge rot\,\vec{n}\right)^2 - \Delta\chi\left(\vec{n}.\vec{H}\right)^2$$

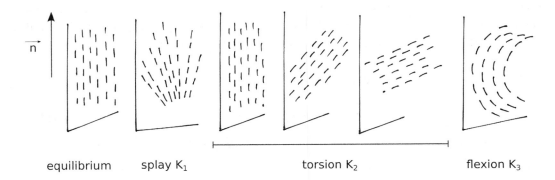

equilibrium      splay $K_1$          torsion $K_2$          flexion $K_3$

FIGURE 5.7 The three types of deformation of nematics.

The three elastic constants are of order $k_B T_c/a$, where $k_B T_c$ is the energy of molecular interaction. The last term is the magnetic energy when a magnetic field is applied. $\Delta\chi = \chi_{//} - \chi_{\perp}$, where $\chi$ is the magnetic susceptibility and $\chi_{//}$, $\chi_{\perp}$, its parallel and perpendicular components, is the anisotropic part of the magnetic susceptibility. If $\Delta\chi$ is positive, which is the case for elongated molecules, the director aligns in the field direction.

From the elastic and magnetic energies one can define three magnetic characteristic lengths:

$$\xi_{H,i} = \sqrt{\frac{K_i}{\Delta\chi}}\frac{1}{H} \text{ with } i = 1,2,3.$$

### 5.2.2 Preparation of Monodomain Samples

By rubbing glass or plexiglass slides against a sheet of drawing paper, Chatelain observed that an orientation of $\vec{n}$ parallel to the axis of friction is imposed [4]; the anchor is called planar (Figure 5.8A). To achieve a "homeotropic" anchorage, where the orientation of the molecules is perpendicular to the surface, the plate is immersed into a surfactant solution (Figure 5.8C). A twisted sample can be prepared when the alignments imposed on the surfaces do not have the same direction (Figure 5.8B). Mono-domain samples are thus achieved, provided that the distance between the two slides is less than 10 μm.

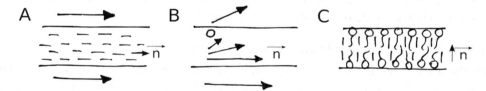

FIGURE 5.8 Different orientations of $\vec{n}$: (A) planar, (B) twisted from $\pi/2$, and (C) homeotropic.

### 5.2.3 Alignment in a Magnetic Field: Fredericks Transition [5]

A nematic lamella of thickness $d$ is aligned parallel to the walls, and we apply a perpendicular magnetic field $\vec{H}$. Then, there is a competition between the alignment imposed by the wall anchorage and the orientation in the direction of $\vec{H}$ (Figure 5.9). A remarkable property is the existence of a threshold field, noted $\vec{H}_C$. If $H < H_C$, the sample remains undistorted because the anchoring forces are dominant and the planar orientation is maintained (Figure 5.9A). If $H > H_C$, the orientation in the field starts to prevail, and the molecules align with the magnetic field in the center of the cell (Figure 5.9B).

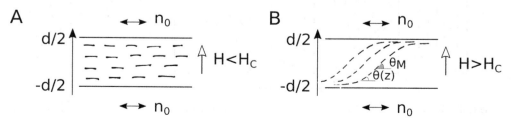

FIGURE 5.9 Fredericks transition: (A) $H < H_c$; (B) $H > H_C$. $\theta(z)$ is the tilt angle. $\theta_M$ is the order parameter of the Fredericks transition, which vanishes at $H_c$.

Between the crossed polarizer and analyzer, the visual field remains dark until the field reaches $H_C$ and suddenly becomes brighter at $H_C$. Before performing exact calculations, we can estimate $H_C$ from the competition between the deformation energy $\approx K/d^2$ (with $d$ the distance between the walls, and $K$ the Franck constant of the deformation mode of interest) and the gain of magnetic energy $\approx \Delta\chi H^2$:

$$H_C = (\pi/d)\cdot\sqrt{(K_i/\Delta\chi)}, \text{ or equivalently the magnetic length } \xi_i \text{ is equal to d.}$$

### 5.2.3.1 Landau Description of the Fredericks Transition: Statics [6]

A nematic monocrystal shown in Figure 5.9 is submitted to a magnetic field $H$ perpendicular to $\vec{n}_0$. If $\theta$ is the angle between $\vec{n}_0$ and $\vec{n}$, the Franck free energy can be written as:

$$F\Big|_{m^2} = \frac{1}{2}\Delta\chi H^2 \int_{-d/2}^{d/2} dz \left[ \xi^2 \left(\frac{d\theta}{dz}\right)^2 - \sin^2\theta \right]. \tag{5.1}$$

For simplicity, we have assumed that $K_1 = K_2 = K_3 = K$, which sets $\xi^2 = (K/\Delta\chi)(1/H^2)$.

In the limit of strong anchoring, the optimal configuration near $H_C$ corresponds to $\theta = \theta_M \cos(\pi z/d)$ and Equation 1 becomes:

$$F\Big|_{m^2} = \frac{d}{2}\Delta\chi H^2 \left[ \frac{1}{2}a\theta_M^2 + \frac{1}{4}b\theta_M^4 \right] \tag{5.2}$$

where $a = (\xi\pi/d)^2 - 1 = 2\left[\dfrac{H_C - H}{H_C}\right]$ and $b = 1/2$, by setting $(d/\xi\pi) = (H/H_C) = h$, which defines $H_C$:

The evolution of $F$ as a function of $\theta_M$ for different values of $h$ is shown in Figure 5.10A. For $h < 1$, the solution is $\theta_M = 0$.

For $h > 1$, $\theta_M^2 = -(a/b) = 4\varepsilon$ where we set $\varepsilon = (H - H_C/H_C)$. We have two solutions $\theta_M$ and $-\theta_M$, which are separated by a wall, as shown by Figure 5.10B,C. For a solution $\theta = \theta(x)\cos\left(\dfrac{\pi z}{d}\right)$, the Franck energy becomes:

$$F\Big|_{m^2} = \frac{d}{2}\Delta\chi H^2 \left[ \frac{1}{2}a\theta_M^2 + \frac{1}{4}b\theta_M^4 + \frac{1}{2}\xi^2 \left(\frac{\partial\theta_M}{\partial x}\right)^2 \right]$$

We set $\xi_\perp = \xi(2\varepsilon)^{-1/2}$. $\xi_\perp$ is the width of the wall between domains (+) and (−), which diverge at $H_C$ (Figure 5.10) and $\theta_M(x) = \theta_M \text{th}(x/2\xi_\perp)$.

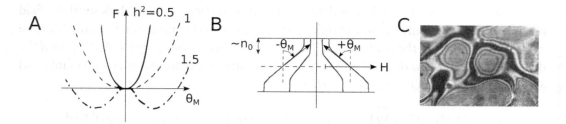

FIGURE 5.10   (A) Free energy of a nematic lamella as a function of the tilt angle $\theta_M$. (B) Wall separating two zones with opposite tilt ($\theta_M$ and $-\theta_M$). (C) Typical aspect of the sample, slightly above $H_C$ between crossed polarizers. (Courtesy of L. Léger.)

### 5.2.3.2 Dynamics of the Fredericks Transition: Critical Slowing Down [7]

The dynamics of the Fredericks transition is an example of the dynamics of phase transition, which is well described by the Van Hove approximation. The friction coefficient associated with the rotation of the director is the rotational viscosity coefficient $\gamma_1$ [1]. The balance between viscous and elastic torques can be written as:

$$\gamma_1 \frac{\partial \theta_M}{\partial t} = -\frac{\partial F}{\partial \theta_M} \tag{5.3}$$

By setting $\tau = \left( \gamma_1 / \Delta\chi H^2 \right)$, Equation 2 and Equation 3 lead to:

$$\tau \frac{\partial \theta_M}{\partial t} = (h^2 - 1)\theta_M - \frac{\theta_M^3}{2} \tag{5.4}$$

This equation shows that the fluctuations of $\theta_M$ slow down at $H = H_C$.

If we consider a small fluctuation of $\theta_M$ around the equilibrium tilt angle $\theta_M^e$ (i.e. $\theta_M = \theta_M^e \pm \delta\theta_M$), the linearized form of Equation 5.4 leads to the solution:

$$\delta\theta_M(t) = \delta\theta_M(t_0) \cdot \exp(-1/\tau_r) \text{ with } \tau_r = \tau(4\varepsilon)^{-1}.$$

## 5.2.4 Alignment in an Electric Field: Display Applications

Twisted nematics (Figure 5.11) are used in almost all common liquid crystal displays. There is no single molecule with the optimal properties for this application, and, as a result, all existing displays up to now use mixtures of 10 to 20 liquid crystals of different structures.

Nematics are highly polarizable and align along an electric field because $\varepsilon_{//} > \varepsilon_\perp$, where $\varepsilon$ is the polarizability.

In the initial state where $\vec{E} = \vec{0}$, the molecules have a twisted arrangement. The polarization $\vec{P}$ of light follows the orientation of $\vec{n}$. Between parallel polarizers, there is a total extinction of the light (Figure 5.11A). On the contrary, as soon as $\vec{E} > \vec{E}_C$ (defined as the distortion field threshold) the sample becomes transparent (Figure 5.11B).

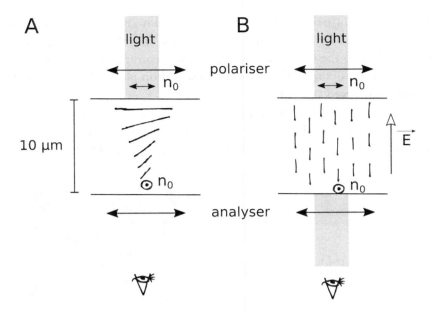

**FIGURE 5.11** Polarization of a nematics. On the left, in the absence of an electric field, the initial state (a twisted nematic) is opaque. On the right, under field $E$, the sample is transparent.

This is the principle of the display system achieved through engraved microelectrodes and colored filters (such as for a watch or a flat screen). Another application is the electrical control of the transparency of a window (Figure 5.12). When the field is cut off, the liquid crystal relaxes to its initial state within 10 to 100 ms depending on the molecular system parameters.

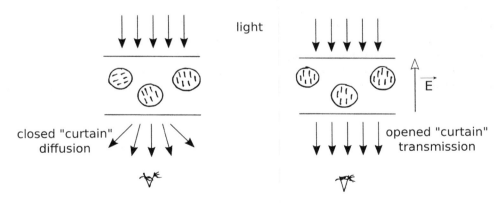

**FIGURE 5.12** Nematic droplets in a window: opaque (disordered alignment) and transparent (alignment along $E$).

### 5.2.5 Nematics Textures

Defects in crystals are called *dislocations*. The defects in liquid crystals are called *disclinations* [8]. The microscopic image of these defects is called a *texture*.

The observation of a nematic liquid crystal between cross-polarizers (Figure 5.13B) shows a texture formed by black filaments which are singular lines of molecular alignment corresponding to the light extinction positions, connected together by nuclei (such as defect points or lines perpendicular to the observation plane). Line-like defects are the more commonly observed structures.

A defect is of rank $s$ if, when moving around a closed path contained in the plane perpendicular to the defect line, the director rotates by a quantity $2\pi s$. Figure 5.13A shows four examples of disclinations corresponding to $s = 1/2$ and $s = -1/2$ at the top and $s = -1$ and 1 at the bottom.

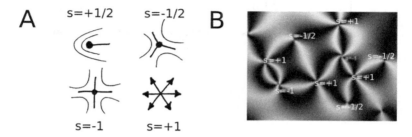

FIGURE 5.13   (A) Four examples of disclinations in nematics. The sticks represent the molecules; the dot represents the defect line perpendicular to the plane of the figure and $s$ refers to the rank of the defect. (B) Experimental texture of nematics where we find the different types of disclinations listed in (A).

## References

1. P.-G. de Gennes, J. Prost, *The Physics of Liquid Crystals*, 2nd Edition. Clarendon Press, 1995.
2. F.C. Franck, *Discuss. Faraday Soc.*, 25, 1 (1958).
3. C.W. Oseen, *Trans. Faraday Soc.*, 29, 883 (1933).
4. P. Chatelain, Bull. *Soc. Franc. Mineral.*, 66, 105 (1943).
5. V. Fredericks, V. Zolina, *Trans. Faraday Soc.*, 29, 919 (1933).
6. P. Pieranski, F. Brochard, E. Guyon, *J. Phys.*, 33, 681 (1972).
7. P. Pieranski, F. Brochard, E. Guyon. *J. Phys. (Paris)* 34, 35 (1973).
8. M. Kleman, *Points, lignes, Parois dans les fluides anisotropes. Tome 1.* Les Editions de Physique, 1977.

# CHAPTER 6

# Surfactants

Surfactants are molecules that lower surface tensions of liquids and interfacial tensions between liquid and solid or between two immiscible liquids. They have some affinity for interfaces because they are amphiphilic molecules.

## 6.1 AMPHIPHILIC MOLECULES

*Amphiphilic molecules* (Figure 6.1) are composed of two antagonistic moieties:

- An oil-loving (lipophilic) tail
- A water-loving (hydrophilic) head

This dual nature imparts an affinity for interfaces to amphiphilic molecules, which has the direct consequence of lowering surface tensions. Hence their name: surfactants. These molecules play a considerable role in the cosmetics, pharmaceutical and chemical industries (paints, detergents).

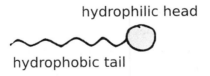

FIGURE 6.1   Schematic representation of an amphiphilic molecule.

The lipophilic part consists of one or two hydrogenated paraffinic chains $CH_3-(CH_2)_n-$, or sometimes fluorinated $CF_3-(CF_2)_n-$, for superhydrophobic treatments (see Section 4.5). The classification of surfactants is based on the chemical nature of the hydrophilic part.

### 6.1.1 Classification of Amphiphilic Molecules

The polar head can be derived from dissociated salts in an aqueous medium. The positive or negative charge provides surfactants with their hydrophilic character. There are also

TABLE 6.1    Classification of Surfactants

| | |
|---|---|
| Anionic | $CO_2^-$ ($Na^+$) fatty acid salt<br>$SO_4^-$ ($Na^+$) sulfate (e.g. sodium dodecyl sulfate (SDS))<br>$SO_3^-$ ($Na^+$) sulfonate (e.g. AOT, bis-(2-ethylhexylsulfosuccinate)) |
| Cationic | $$R - \overset{\displaystyle R}{\underset{\displaystyle R}{NH_4^+}} - R$$<br>Amine or quaternary ammonium salts |
| Non-ionic | $R \left( CH_2 - CH_2 - O \right)_N$<br>Polyoxyethylene (POE) (also called PEG, polyethylene glycol) |
| Amphoteric<br>(anion + cation) | $R - COO - CH_2$<br>$\mid$<br>$R - COO - CH$<br>$\mid$<br>$CH_2 )$<br>Choline + phosphate group<br>(e.g. lecithin from egg yolk) |

non-ionic polar parts, such as polyoxyethylene (water-soluble polymer), and globally electrically neutral parts, such as amphoteric groups (Table 6.1). Phospholipids, which are the main components of biological membranes, are of the latter type; they self-assemble into bilayers and form uni- or multi-lamellar vesicles.

## 6.1.2  Roles of Surfactants at Interfaces

In general, surfactants are positioned at interfaces and lower the interfacial energy.

At the surface of a (polar) liquid, they spontaneously form a monolayer or monomolecular film (Figure 6.2.). Surface tension decreases: $\gamma = \gamma_0 - \Pi$, where $\gamma_0$ is the surface tension of the pure liquid and $\Pi$ is the surface pressure exerted by the amphiphilic molecules.

Surfactants also adsorb onto solid surfaces and alter their wettability (Sections 4.5 and 4.6). This phenomenon has important applications in the fields of lubrication, mineral flotation, and dispersion of colloidal suspensions (Figure 6.2.).

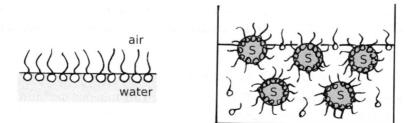

FIGURE 6.2   On the left, surfactants at the surface of a liquid. On the right, surfactants adsorbed on solids: application to the extraction of minerals by flotation.

### 6.1.3 Self-Assembly and Aggregation in Water

In water, surfactants aggregate into micelles. These are structures typically composed of about a hundred surfactant molecules. Micelles form when the surfactant concentration exceeds the *CMC* or critical micellar concentration (Figure 6.3.). This micellar structure reduces the contact area between lipophilic chains and water, thus decreasing the energy of the system.

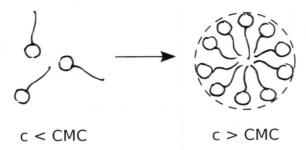

c < CMC          c > CMC

FIGURE 6.3    Micelle formation at CMC.

Let us consider the chemical equilibrium between surfactants and micelles as shown in Figure 6.3. At equilibrium, the chemical potentials (noted μ) of the molecules of free (or monomeric) surfactants (at concentration $x_1$) and surfactants engaged in a micelle (at concentration $x_n$) are equal. We have:

$$\mu = \mu_1 + kT \cdot \ell n(x_1) = \mu_n + \left(\frac{kT}{n}\right) \cdot \ell n\left(\frac{x_n}{n}\right).$$

Moreover, the conservation law is written:

$$x_1 + x_n = c,$$

with $c$ the total concentration of surfactants.

From the equality of chemical potentials, we obtain:

$$\frac{x_n}{n} = \left(\frac{x_1}{CMC}\right)^n,$$

where *CMC* is defined by: $CMC = e^{-(\mu_1 - \mu_n)/kT}$

If $c < CMC$, $x_n = 0$.

If $c > CMC$, $x_1 = CMC$ and $x_n = c - CMC$. For $c > CMC$, the concentration of free surfactants remains constant. The surface tension becomes independent of the concentration of surfactants (see Figure 6.12), which leads to a measure of the *CMC*.

The hydrophobic core of the micelles also serves to dissolve hydrophobic solutes in water. At very high concentrations, surfactants form lyotropic liquid crystals (Section 5.1 and Figure 6.4.).

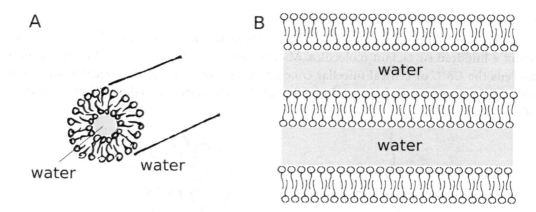

FIGURE 6.4 Morphology of aggregates formed by amphiphilic molecules in aqueous solution. (A) cylindrical phase and (B) lamellar phase.

### 6.1.4 Water/Oil Emulsions and Hydrophilic–Lipophilic Balance

When water and oil are stirred, droplets of oil in water or water in oil can be formed. But these emulsions are not stable. Oil and water eventually segregate and form two layers (the less dense oil on top of the water surface). To stabilize oil/water and water/oil emulsions, a surfactant must be added. More than 15,000 amphiphilic molecules have been listed. The choice of the appropriate surfactant is guided by the hydrophilic–hydrophobic balance (HLB). This empirical methodology was introduced by Griffin (1949) and provides a measure of the hydrophobic–hydrophilic ratio for each surfactant on a range from 0 to 20 (Figure 6.5.).

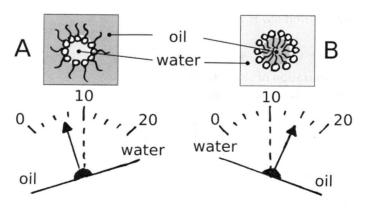

FIGURE 6.5 The hydrophilic–hydrophobic balance (HLB): (A) drop of water in oil; (B) drop of oil in water.

To emulsify oil in water, hydrophilic dominant surfactants, i.e. with HLB > 10 (Figure 6.6. – e.g. oleic acid HLB = 17, pine oil HLB = 16), should be used. For reverse emulsions, surfactants of HLB < 10 would be preferred (Figure 6.6 – e.g. cocoa butter HLB = 6, soybean oil HLB = 6). If two surfactants are mixed, the HLB of the mixture is the sum of the HLBs of each amphiphilic molecule weighted by their fraction.

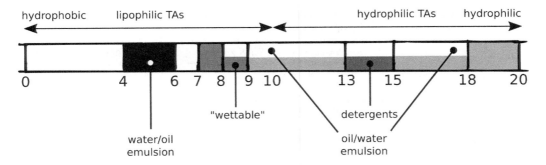

FIGURE 6.6   HLB scale.

## 6.2 MONOMOLECULAR FILMS OF AMPHIPHILIC MOLECULES

In 1774, Benjamin Franklin poured a spoonful of oil (~1 ml) onto a pond surface, which was wrinkled by a light breeze and became as smooth as a mirror. The film that was formed at the surface covered about 400 m². Knowing the volume of the spoonful of oil, Franklin calculated that the film had a thickness of only 2.5 nm, i.e. a molecular size.

Between 1917 and 1940, Irving Langmuir studied the structure of surfactant films on the water surface. For this purpose, he created a device that was named after him, the Langmuir film balance. In the 1930s, with Katharine Blodgett, he deposited successive monolayers on glass substrates, thus creating the first anti-reflective lenses (Section 9.1).

In 1972, J. Michaël Kosterlitz and David J. Thouless (Nobel Prize in Physics 2016) developed a theory predicting a new type of phase transition specific to two-dimensional systems. While their theory has found applications in superconductors and superfluidity, Langmuir films quickly emerged as an ideal 2D system to study the physics of these phase transitions. Thanks to the use of intense X-ray sources, crystallography on Langmuir monolayer has been achieved, leading to a deep knowledge of this 2D physics rather than to spectacular applications in everyday life.

### 6.2.1 Insoluble Films

Alcohols and hydrophilic fatty acids with long water-insoluble chains (e.g. stearic acid: $C_{17}H_{35}COOH$) are used.

To form a so-called Langmuir film, the surfactant is dissolved in a volatile solvent (e.g. chloroform, methanol) and deposited drop by drop at the water–air interface. The solvent evaporates. The presence of the hydrophilic head allows anchoring to the water surface and spreading of the surfactant, which forms a layer of monomolecular thickness to minimize the surface energy. The use of alkanes (e.g. hexadecane $C_{16}H_{34}$) leads to the formation of lenses on the water surface, allowing hydrophobic chains to minimize their contact with water (Table 6.2).

TABLE 6.2 Example of Surfactants and Importance of the Polar Head Allowing the Formation of Monomolecular Layers

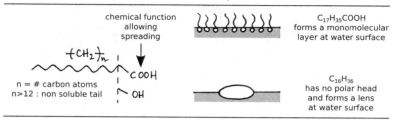

### 6.2.1.1 Isotherms Π(A)

In a so-called Langmuir trough equipped with barriers generally made of Teflon that move on the surface of the water, the area of the monolayer formed at the interface can be varied. As the area of the monolayer decreases, the surface density of surfactants, 1/A (where A is the projected area per polar head, in the order of a few Å$^2$) increases and the surface tension decreases. This variation in interfacial tension is noted $\Pi = \gamma_0 - \gamma$ (Sections 4.1 and 4.5). $\Pi$ can also be seen as the surface pressure exerted onto the barriers by the compressed monolayer (Figure 6.7). The Wilhelmy plate method allows the measurement of $\Pi$. It is a very clean (and therefore hydrophilic) thin blade (made of platinum or paper soaked in water) partially immersed in the Langmuir trough and connected to a microbalance. The capillary force exerted at the contact line balances the weight and Archimedes' buoyant force, which allows us to derive $\gamma = $ (Weight − Buoyant force)/Wilhelmy plate's perimeter. The hydrophilic nature of the Wilhelmy plate ensures a zero-contact angle. The direction of the capillary force is vertical.

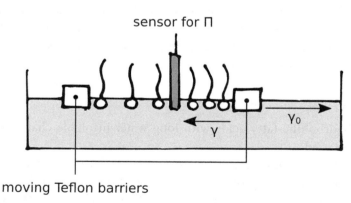

FIGURE 6.7 Principle of the Langmuir trough.

If the temperature is fixed, the isotherms $\Pi(A)$ are the two-dimensional analogs of the curves $P(V)$ in three dimensions (with $P$ the pressure and $V$ the volume).

A compression isotherm $\Pi(A)$ is generally characterized by regions where the pressure increases as the available area per amphiphilic molecule decreases and by regions marked by a plateau of $\Pi$. The pressure plateaus correspond to transitions between different phases, which find an equivalent in 3D (Figure 6.8): the gas phase, the liquid expanded phase (or 2D liquid), liquid condensed (LC) (or 2D liquid crystal), and solid (or 2D crystal).

State of the film

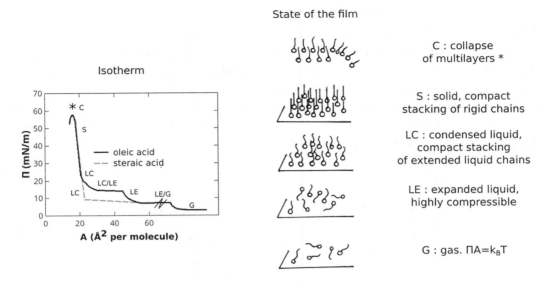

Isotherm

C : collapse
of multilayers *

S : solid, compact
stacking of rigid chains

LC : condensed liquid,
compact stacking
of extended liquid chains

LE : expanded liquid,
highly compressible

G : gas. $\Pi A = k_B T$

FIGURE 6.8   Isotherm $\Pi(A)$ and description of the different states of the monomolecular film.

Although the thickness of Langmuir monolayers is nanometric, it is easy to observe these phase transitions with an optical fluorescence microscope (Figure 6.9) by adding a small fraction of amphiphilic molecules carrying a fluorescent group (Figure 6.9). Because the fluorescent surfactant is more bulky, it is generally excluded from the LC and S phases and is incorporated into the LE phase (Figure 6.9). In the G phase, its density is too low to be detected.

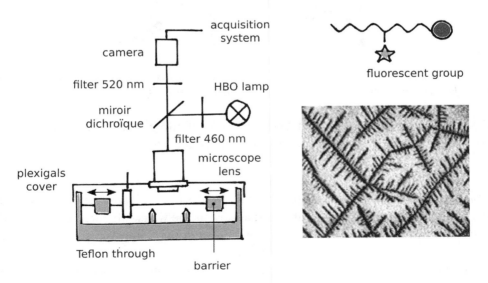

FIGURE 6.9   Drawing of a fluorescence microscope combined with a Langmuir trough and balance. Schematic representation of a fluorescent surfactant. Fluorescence microscopy image of dendritic (dark) LC domains in a homogeneous (bright) LE phase.

### 6.2.1.2 Langmuir–Blodgett (LB) Films

LB films are monolayers of surfactant molecules initially formed at the air–water interface and transferred to a solid support. Starting from a submerged hydrophilic solid substrate, a monomolecular film is formed when the solid is withdrawn (Figure 6.10). By repeated dipping, several layers can be transferred (LB film). The kinetics (i.e. the speed of withdrawal of the support) are important. At high speed $U > U_M$ (where $U_M$ is the critical speed of forced wetting as discussed in Chapter 4), water is dragged (forced wetting), and defects are generated within the film (Figure 6.11).

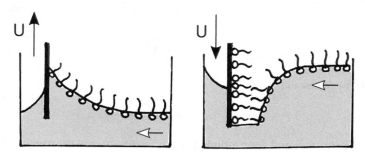

FIGURE 6.10   Formation principle of LB films.

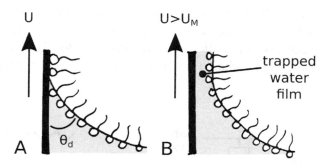

FIGURE 6.11   Dynamics of film deposition: (A) $U < U_M$: optimal conditions – defect-free deposition. (B) $U > U_M$: an intercalated water film is dragged with the substrate.

Anti-reflective glass (K. Blodgett, Section 9.1), antifouling coating, and molecular electronics [1] are among the most common applications of LB films.

## 6.2.2 Soluble Films

Soluble films are made of short amphiphilic molecules ($n < 12$), and they are called Gibbs films. Surfactants are soluble in water and adsorb at the water–air interface. When their volume concentration increases, the surface tension of the water decreases and reaches a plateau. From this concentration, called the critical micellar concentration (CMC, Section 6.1), the interface is saturated, and the solubilized surfactants form micelles (Figure 5.12).

In water at room temperature, sodium dodecyl sulfate, $CH_3(CH_2)_{11}OSO_3^-Na^+$, anionic (SDS) has a CMC of 8.5 mmol $L^{-1}$; dodecyltrimethylammonium bromide, $C_{12}H_{25}(CH_3)_3NBr$, cationic (DTAB) has a CMC of 15.3 mmol $L^{-1}$ (Figure 6.12).

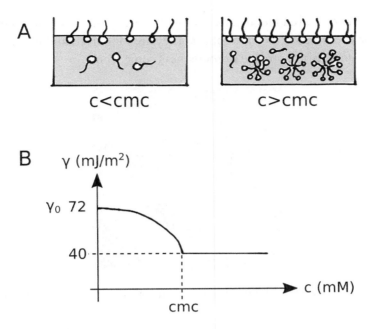

FIGURE 6.12 (A) Formation of micelles for $c > CMC$. (B) Gibbs isotherm.

Reference

1. S.A. Hussain, *Mod. Phys. Lett. B*, 23, 1–15 (2009).

## 6.3 SOAP FILMS – BUBBLES AND VESICLES

Since antiquity, bubbles (Figure 6.13) have fascinated poets and scientists because they are fragile and short-lived. You blow on a soap film whose thickness is in the order of 100 nm, and a colorful bubble flies away and bursts.

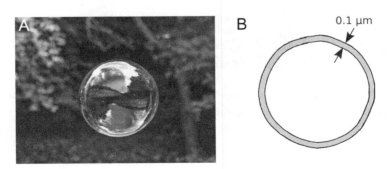

FIGURE 6.13 **(See color insert.)** (A) Soap bubble. (Copyright Shutterstock.) (B) Its membrane has a thickness of about 0.1 μm.

Vesicles (Figure 6.14), red blood cells, living cells are made of a lipid membrane whose thickness is about 6 nm, but they are robust. They are long-lived and do not burst spontaneously. Where does this difference in robustness come from?

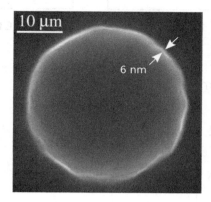

FIGURE 6.14    Lipid vesicle about 30 μm in diameter. Its membrane has a thickness of about 6 nm.

### 6.3.1 Soap Films, Bubbles, and Foams

#### 6.3.1.1 Films

When shined with a monochromatic light, interference fringes are visible on the soap film. Under white light illumination, the wavelength of the reflected light corresponds to the interference condition (Figure 6.15 and 6.16).

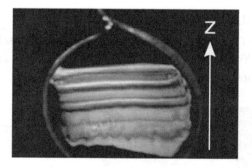

FIGURE 6.15    **(See color insert.)** Vertical soap film obtained by drainage, with a thickness $e(z)$ varying along the vertical axis. At the top the film is black.

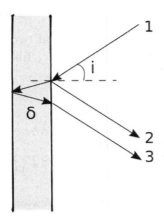

FIGURE 6.16   Interferences created from the light rays reflected at the two interfaces of a soap film and coming from an incident beam (with an angle of incidence $i$).

Under normal incidence ($i \approx 0$), the incident beam 1 of wavelength $\lambda$ is reflected at two interfaces and gives rise to two reflected beams 2 and 3. The phase shift $\Delta\varphi$ between 2 and 3 is given by:

$$\Delta\varphi = 2\pi\frac{\delta}{\lambda} + \pi,$$

where $\delta$ is the difference in optical path ($\delta$ = distance × refractive index of the medium – here $n = 1.33$ for water). The additional phase shift $\pi$ is the phase inversion that occurs at the interface when passing from low index (air $n = 1$) to a high index medium.

$$\delta = 2ne$$

If the thickness $e$ is very small compared to $\lambda$ (about 0.5 μm for visible light), $\Delta\varphi \approx \pi$, interferences are destructive, and the film is black.

If $\delta = k\lambda + 1/2$ (with $k$ being an integer), interferences are constructive, which corresponds to a maximum intensity.

### 6.3.1.2 Foams

Foams are formed from the dispersion of a large volume of gas in a small volume of liquid containing surfactants that adsorb at the gas–liquid interface of the bubbles.

There are two categories of foams:

- *Liquid foams*: Beer made with alcohol, carbon dioxide, and polypeptides, snow eggs made from albumin, carbon foams contained in fire extinguishers,

- *Solid foams*: Meringues, expanded polymers, volcanic rocks.

### 6.3.1.3 Stability of Soap Films

Isaac Newton (1704) studied soap films. He found two states of equilibrium: the black film ("Newton black film" (NBF)) and the thicker film ("Common black film" (CBF)) (Figure 6.17).

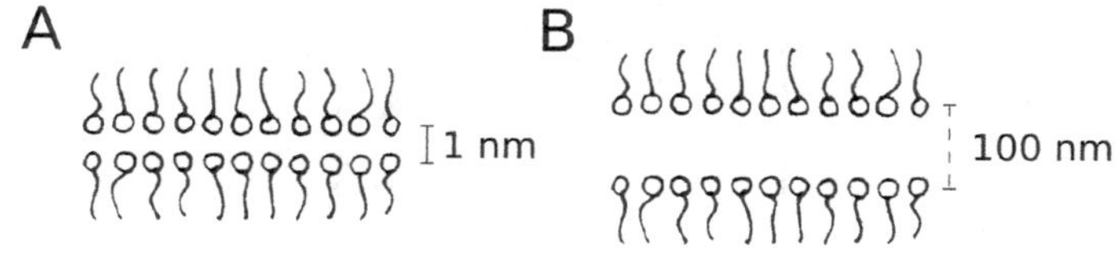

A          B

FIGURE 6.17    Two equilibrium states of soap films: (A) Newton Black Film. (B) Common Black Film.

Thicknesses of NBF and CBF are given by the minima of the film energy $F(e)$. The energy of the film is the sum of the surface energy and the contribution of the long-range forces $P(e)$:

$$F(e) = 2\gamma + P(e),$$

where $e$ is the film thickness, and $P(e) \rightarrow 0$ when $e \rightarrow \infty$.

If surfactants are charged: $P(e) = -\dfrac{A_{LL}}{12\pi e^2} + B\exp(-\kappa_d e)$, with $A_{LL}$ the Hamaker constant, $\kappa_d^{-1}$ the Debye length (Sections 2.4, 2.5, 4.4, 4.5), and $B$ a constant.

The shape of the curve $F(e)$ is shown in Figure 6.18. The two minima of $F(e)$ correspond to NBF and CBF. In the presence of salt, electrostatic interactions are screened, and the secondary minimum disappears.

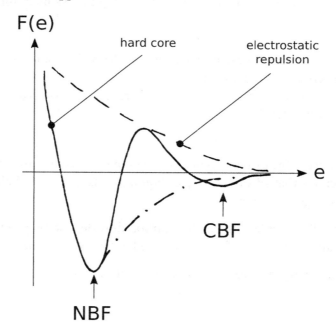

FIGURE 6.18    Shape of the curve $F(e)$ expressing the energy of the soap film as a function of its thickness.

### 6.3.1.4 Bursting of Films and Bubbles

After soap bubbles are formed, they undergo drainage, get thinner, and ultimately burst. A hole is created in the soap bubble, very often because of dust, then it grows and is surrounded by a rim that collects the liquid from the hole (Figure 6.19). The equation that expresses hole opening is given by the fundamental equation of mechanics applied to the rim of mass $M$:

$$f = \frac{dP}{dt}$$

where $P = MV$ is the transfer momentum per unit length (along the circumference of the hole) of the rim of mass $M$.

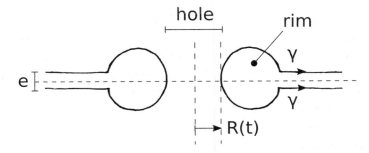

FIGURE 6.19  Formation and propagation of a hole in a soap bubble.

$M$ increases during the opening of the hole because $M = \rho Re$ with $\rho$ the density of the water, $R$ the radius of the hole, and $e$ the thickness of the rim. If we make the hypothesis (which has been experimentally tested) that the velocity $V$ is constant (or hardly varies compared to the mass accumulation in the rim), then: $dP/dt = \rho e(dR/dt)V = \rho eV^2$.

The force $f$ per unit length that drives film opening is the surface tension:

$f = 2\gamma$ (the factor 2 comes from the fact that the surface tension is applied on both sides of the film). So we get:

$$2\gamma = \rho eV^2 \text{ i.e. } V = \left(2\gamma/\rho e\right)^{1/2}$$

This velocity is high, typically in the order of 10 m s$^{-1}$.

### 6.3.2 Lipid Bilayers: Vesicles and Living Cells

### 6.3.2.1 Vesicles

Vesicles are made of a lipid bilayer, separating the internal and external medium. Small Unilamellar Vesicles (SUVs), with a diameter of ~100 nm, are used in the cosmetics and pharmaceutical industries for drug transfer and membrane regeneration (AIDS, burns). Giant Unilamellar Vesicles (GUVs) have sizes comparable to those of cells (~1 to 50 µm). They are formed by "swelling" a lamellar phase of phospholipids or by impinging a jet onto a lipid bilayer. They are used as a model system for cell membranes (Figure 6.20).

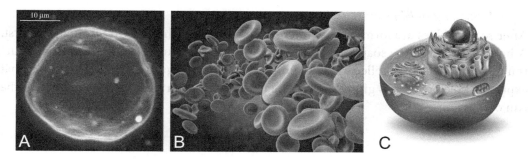

FIGURE 6.20 Vesicles and living cells: (A) GUV. (B) Red cells. (C) Artistic representation of a cell. (B,C copyright Shutterstock.)

### 6.3.2.2 Transient Pores in Tense Vesicles and Nuclear Membranes

Phospholipids are insoluble. At equilibrium, the vesicle minimizes the surface area $A$ per polar head, which means that $\partial F/\partial A = 0$, or equivalently $\sigma = 0$. In fact, there is still a small residual tension ($\sigma \sim 10^{-4}$ mN m$^{-1}$). Vesicles can be inflated either by osmotic shock or by illumination. But tense vesicles will not necessarily burst. An extraordinary phenomenon may happen. A pore is created and opens, then closes again (Figure 6.21). The force that opens the hole is $2\sigma$. Unlike soap film, the tension of the membrane depends on swelling, as with a balloon that is blown. As soon as the hole appears and grows, tension drops, and the pore stops growing. Due to line tension, the hole will close again while the internal liquid leaks to maintain the membrane tension close to zero. Both vesicles and red blood cells are self-healing systems (Figure 6.22). Even when inflated, they do not burst, which explains their robustness.

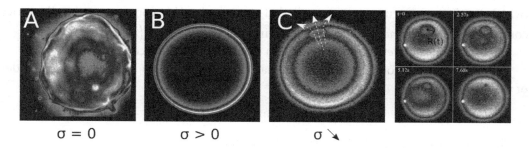

$\sigma = 0$      $\sigma > 0$      $\sigma \searrow$

FIGURE 6.21 **(See color insert.)** Transient pore: (A) Floppy vesicle, membrane tension (almost) zero. (B) Tense vesicle after light irradiation. (C) Opening of a pore: $R(t)$ increases. Sequence of four images: pore closure: $R(t)$ decreases. The time is indicated in seconds. (Extracted from [1].)

FIGURE 6.22 Self-healing process of a lipid membrane.

The nucleus of a cell is also surrounded by a nuclear membrane that protects the chromosomes and filters the passage of specific molecules. When cells sneak into the extracellular matrix, they deform, and the nucleus membrane becomes porous: transient pores open and close, leaving genetic material behind. It can even explode in extreme cases, resulting in the release of DNA and cell death.

Reference
1. O. Sandre, Pores transitoires, adhésion et fusion des vésicules géantes. PhD thesis, Université Paris 6 (2000).

# CHAPTER 7

# Polymers

## 7.1 POLYMERS: GIANT MOLECULES

A polymer is a long chain obtained from the repetition of small "monomer" units joined by covalent bonding. The degree of polymerization (or number of monomeric units) N is in the range of $10^3$ to $10^5$. If nature is able to create chromosomes made up of $10^8$ bases long DNA molecules (Section 10.1), it is already quite an achievement to chemically create polymers made of $10^5$ monomers without error.

Polymers have become key materials. They are found everywhere in everyday life (plastics, glues, tires, textiles, paints). Figure 7.1 gives some examples of synthetic polymers and typical applications.

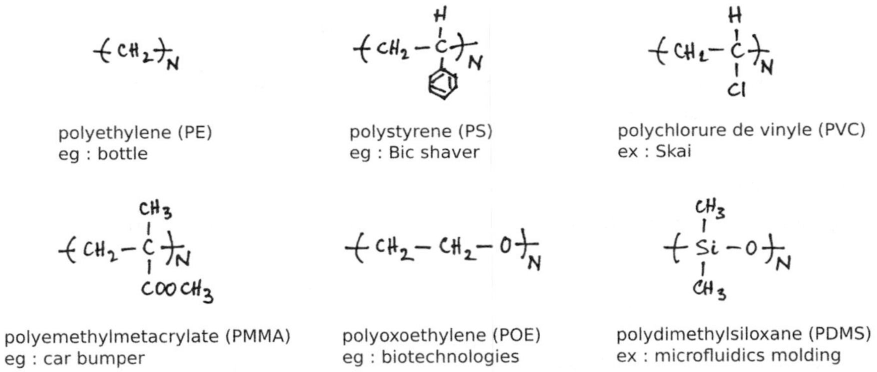

$$\left( CH_2 \right)_N$$

polyethylene (PE)
eg : bottle

$$\left( CH_2 - \overset{\underset{H}{|}}{\underset{\bigcirc}{C}} \right)_N$$

polystyrene (PS)
eg : Bic shaver

$$\left( CH_2 - \overset{\underset{H}{|}}{\underset{Cl}{C}} \right)_N$$

polychlorure de vinyle (PVC)
ex : Skai

$$\left( CH_2 - \overset{\underset{CH_3}{|}}{\underset{COOCH_3}{C}} \right)_N$$

polyemethylmetacrylate (PMMA)
eg : car bumper

$$\left( CH_2 - CH_2 - O \right)_N$$

polyoxoethylene (POE)
eg : biotechnologies

$$\left( \overset{\underset{CH_3}{|}}{\underset{CH_3}{Si}} - O \right)_N$$

polydimethylsiloxane (PDMS)
ex : microfluidics molding

FIGURE 7.1 Main synthetic polymers and applications.

### 7.1.1 Chemical Synthesis

Natural polymers (silk, cotton, wood, etc.) have been used for thousands of years by our ancestors. The first industrial process using chemically modified polymers was the invention of rubber by Goodyear in 1839, followed by the chemical modification of cellulose (1865). But, at this time, polymer science was stagnating because the concept of polymer was not understood. It was only in 1920 that the chemist Staudinger was able to demonstrate the existence of long chain systems called "polymers." Polymer chemistry then started to bloom.

There are three main processes for the synthesis of long chains, and they are based on two polymerization principles: (1) by successive addition of monomers according to the $A_n + A$ gives $A_{n+1}$ rule, or (2) by condensation of chains at the reactive ends according to the $A_n + A_m$ gives $A_{n+m}$ rule.

### 7.1.1.1 Free Radical Polymerization

Chains grow continuously by opening a double bond at each step.

*Initiation*: the reaction is initiated using a free radical $R°$ ("broken" molecule obtained by UV radiation or high temperature), which is neutral but possesses an unpaired electron (e.g. $CH_3°$). This highly reactive species can be associated with a monomer $M$.

*Propagation*

$$R° + M \rightarrow RM°$$

$$RM° + M \rightarrow RM_2°$$

*Termination*

$$RM_p° + RM_q° \rightarrow RM_{p+q}R$$

Termination of the reaction occurs by combining two radicals. This is a random process; the chains will be polydisperse.

### 7.1.1.2 Anionic or Cationic Polymerization

This synthesis is very similar to the previous one, but the reaction is initiated by ions.

Unlike radical synthesis, there is no possible recombination of cations (or anions). Growth is blocked by the addition of a chemical agent that suppresses the charge. Since all chains grow at the same rate, they will have the same size when arrest is triggered.

### 7.1.1.3 Condensation Polymerization

This is a reaction between two molecules with two terminal reactive groups, which associate in long chains.

*Polyesters*: reactive groups COOH and OH

*Polyamides* (e.g. nylon): reactive groups COOH and NH$_2$

$$\text{H}_2\text{N}\text{-}(\text{CH}_2)_6\text{-} \text{NH}\overline{|\text{H} \quad\quad \text{HO}|}\text{OC}\text{-}(\text{CH}_2)_4\text{-} \text{COOH}$$

*Note*: the reverse hydrolysis reaction that would sever the chains does not occur because of the presence of hydrophobic aliphatic groups.

If the molecule contains two reactive groups, the reaction yields polymer chains that are in a liquid state at high temperatures (molten polymer), and the polymer is reusable. These are thermoplastic polymers.

If the molecule contains three reactive groups, the polymer grows in three dimensions (branched polymers). It is the family of thermoset polymers. In this book only the properties of thermoplastics will be considered.

### 7.1.2 Polydispersity

The size distribution of polymer chains is characterized by $M_w/M_n$ where $M_n$ is the *number average molecular weight* and $M_w$ the *weight average molecular weight*:

$$M_n = \frac{\sum n_x M_x}{\sum n_x} \quad \text{and} \quad M_w = \frac{\sum n_x M_x^2}{\sum n_x M_x}$$

where $M_x = x M_0$ ($M_0$ the monomer molecular weight) and $n_x$ is the number of molecules of degree of polymerization $(DP) = x$.

For a perfectly monodisperse polymer: $M_w/M_n = 1$.

Radical and condensation syntheses lead to $M_w/M_n \approx 2$: this is the signature of a high polydispersity, because the termination is random. But cationic or anionic synthesis leads to $\left(M_w/M_n\right) \approx 1.01-1.1$. Polydispersity is very weak, but cost is high.

### 7.1.3 Stereochemical Disorder

Polyethylene (PE) exists as a single species. But if the monomer is chiral, as for polystyrene (PS), two states, namely *d* and *g*, can be defined. The chain can grow randomly with a disordered succession of *d,g* (*dgdddgggg* ...), and the polymer is called atactic. If the chain grows in an ordered fashion (*ddddddddddd* ...), the polymer is called isotactic. Finally, if growth is ordered in an alternate fashion (*dgdgdgdgdgdg* ...), the polymer is called syndiotactic. The syntheses described here yield atactic polymers. A process invented by Ziegler and Netta in 1954 makes it possible to control the stereochemical order. Only isotactic and syndiotactic polymers can crystallize, the other ones give rise to transparent amorphous glasses.

### 7.1.4 Glass Transition Temperature

A crystal at low temperature melts at a well-defined melting temperature $T_f$. For polymers, the low temperature state is vitreous. Heating transforms this solid state into a molten polymer state over a wide temperature range $T_g$ (Table 7.1). The glass transition temperature $T_g$ depends on the cooling rate.

TABLE 7.1    Glass Transition Temperature $T_g$ for a Few Polymers

| Polymer | Solid | | Liquid | |
| --- | --- | --- | --- | --- |
| | PDMS | Polybutylene (PB) | PVC | PS |
| $T_g$ (°C) | −123 | −23 | 81 | 100 |

A well-defined transition is usually a signature of crystalline purity. In the 19th century, the absence of a well-defined transition for polymers led dogmatic chemists to throw polymers down the drain and not recognize them as pure compounds. By synthesizing increasingly long chains, Staudinger showed that oligomers behave as pure compounds, with a well-defined temperature $T_f$, and that this property is lost when the chains become longer.

### 7.1.5 Polymer Classes

#### 7.1.5.1 Copolymers

These are polymers resulting from the growth of chains with at least two chemically different types of monomers, succeeding each other in a disordered or regular fashion, such as AAAABBAAABBBBBBB or ABABABABA ...

#### 7.1.5.2 Branched or Star Polymers

Branched polymers have at least one skeleton to which side chains are grafted. This can lead to various configurations as shown in Figure 7.2.

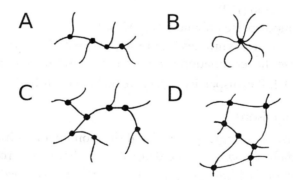

FIGURE 7.2    Polymer A: comb; B: star; C: connected; D: cross-linked.

#### 7.1.5.3 Polyelectrolytes

Polyelectrolytes are polymers made of ionizable monomers that become charged by losing low molecular weight ions called counter-ions, such as an $H^+$ proton in the example shown below:

$$-CH_2-\underset{\underset{COOH}{|}}{CH}-\qquad -CH_2-\underset{\underset{COO^-}{|}}{CH}-$$

Polyacrylic acid (PAA), polystyrene sulfonate (PSS), biopolymers (DNA, alginate – Sections 10.1 and 8.2) are classic examples of polyelectrolytes.

## 7.1.6 Application: Millefeuilles of Polyelectrolytes

A millefeuille is a pastry made up of an alternation of layers of puff pastry and custard cream. A molecular millefeuille is a supramolecular assembly consisting of an alternation of layers of cationic and anionic polyelectrolytes (Figure 7.3). Millefeuilles or multilayers of polyelectrolytes were invented in 1997 at the Charles Sadron Institute in Strasbourg by Gero Decher [1]. They have generated a great interest both from a fundamental and applied point of view. We will discuss their formation, properties, and some applications.

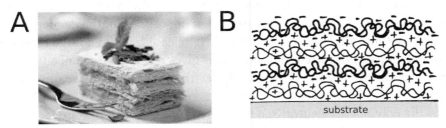

FIGURE 7.3   (A) A millefeuille. (Copyright Shutterstock.) (B) Drawing of a molecular millefeuille.

### 7.1.6.1 Formation of Polyelectrolyte Multilayers: Mechanism and Achievements

At first glance, one might think that there is nothing particularly ingenious about this idea, since we know that there is a Coulomb attraction between positively and negatively charged species. However, by reasoning in this way, we forget an important point: the principle of electroneutrality. Let us consider a positively charged substrate. If this surface is dipped into an anionic polyelectrolyte solution (Figure 7.4), the polymer is expected to adsorb to it via electrostatic interactions. But which process will limit the growth of this layer? Contrary to what one might naively think, it is not that all positive surface charges are neutralized by the anionic groups of the polymer. If this were the case, the second stage of construction of the millefeuille could not be carried out because the new surface has become neutral. In reality, there is *reversal or overcompensation of the charge*.

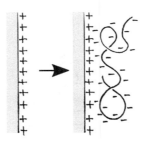

FIGURE 7.4   Adsorption of an anionic polyelectrolyte on a positively charged surface.

From a general point of view, the thickness of the adsorbed layer results from a competition between the attractive electrostatic interactions between polymer and substrate and the repellent steric force due to the entropic elasticity of the polymer chain. But in this picture, the role of counter-ions also becomes crucial. For highly charged polyelectrolytes, above a certain linear charge density $\lambda_c \sim 1/\ell_B$ (with the $\ell_B$ length of Bjerrum, in the order of 7 Å in water at room temperature), counter-ions condense on the polymer. The polyelectrolyte, with its condensed counter-ions, therefore has an effective charge lower than its nominal charge. This phenomenon is called Manning condensation [2]. When the chain adsorbs on a charged surface of opposite sign, the condensed counterions are released into the solution, because their release leads to a large increase in the (translational) entropy of the counter-ions. This gives rise to an uncompensated increase in the polymer charge. For less charged polyelectrolytes, the effect remains without having to invoke Manning condensation, because, for entropic reasons, not all the charges carried by the polymer are neutralized by intimate contact with the surface (as shown in Figure 7.4, the polymer forms loops that extend towards the interface with water or air). It should be noted that this does not mean that the principle of electroneutrality is violated. These residual charges carried by the polyelectrolyte are shielded by counter-ions. However, during the deposition of the second layer, these counter-ions are replaced by polyelectrolyte groups of the opposite sign for the same entropic reasons. This is one of the arguments, given "with the hands" to explain that a charge reversal is possible for the construction of the molecular scaffolding.

How are these molecular millefeuilles produced in practice? Their success is largely due to the simplicity of the formation process. In general, they are prepared by a *Layer by Layer* deposition method (LbL). Initially, Decher [1] proposed the method of substrate soaking, successively in the different polyelectrolyte solutions, separated by rinsing steps (Figure 7.5A). Then, spin-coating (or spin-assisted spreading – Figure 7.5B) and spray (Figure 7.5C) protocols were proposed and validated.

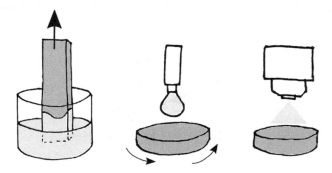

FIGURE 7.5 Schematic representation of the main processes used to manufacture polyelectrolyte multilayers: (A) soaking; (B) spinning; (C) spray. (Adapted from [3].)

### 7.1.6.2 Applications

As seen throughout this book, many of the interactions of an object with its environment depend directly on the nature of the interface. By modifying the surface by adsorption of a layer composed of a different material, it is possible to create materials with new properties. Anti-reflective coatings (Section 9.1) or scratch-resistant coatings are commonly used on our glasses. Particular attention is also paid to very thin films that change the apparent color of the object due to light interference (Section 9.4). This also allows the chemical reactivity of the surface to be modified without affecting the other structural properties of the material, with applications in the biomedical field (implants, prostheses).

The main methods used to deposit organic molecules are the Langmuir–Blodgett method (Section 6.2 and 9.1) or silanization to make hydrophobic surfaces (Sections 8.1 and 4.5). However, these two techniques are complicated to implement: the arsenal of compatible molecules (surfactants or silanes) and the nature of the surfaces (glass-like) are quite limited. On the other hand, the chemistry of polymers is very diverse, which gives access to an almost unlimited chemical library. Polyelectrolyte millefeuilles are thus very convenient and versatile surface coatings with adjustable properties. In addition to surface staining (Figure 7.6A), the mechanical properties of the molecular coating can be modified on demand. This can be done by tuning the chemical nature of polyelectrolytes, inserting metallic nanoparticles between the different layers, or crosslinking the layers with chemical agents to a controlled extent.

FIGURE 7.6  (A) Silicon wafers coated with polyelectrolyte of variable thickness (i.e. a variable number of layers) have different colors. Extracted from www.chem.fsu.edu/multilayers/. (B) Actuator composed of 30 layers of PAA/PAH and one layer of aluminum (seen in electron microscopy) that bends under the effect of humidity. (Adapted from [3].) (C) Scanning electron microscopy images of red blood cells in their discocyte and echinocyte shape covered with ten layers of polyelectrolytes. (Extracted from [4].)

By using polymers that are responsive to the environment (light, pH, etc.), "smart" or functional coatings can be envisioned. By depositing a thin aluminum film (150 nm thick) on a 2 μm thick multilayer polyacrylamide/polyacrylic acid (PAA/PAH) membrane, a Chinese group produced a moisture-sensitive actuator (Figure 7.6B). While the aluminum layer is inert to moisture, the PAA/PAH assembly swells by capturing moisture from the air. This bimorph effect causes the actuator to bend. When the humidity in the air decreases, the multilayer shrinks, and the device returns to its original shape.

Finally, the use of polyelectrolyte multilayers is not limited to the coating of flat surfaces. Professor Möhwald's group at the Max Planck Institute in Potsdam used red blood cells as a template to form polyelectrolyte capsules after successive soaking and then oxidative digestion of the hematocyte (Figure 7.6C). Many strategies for encapsulation and controlled release of active species for medical applications are then possible.

References
1. G. Decher, *Science*, 277, 1232–1237 (1997).
2. G. Manning, *J. Chem. Phys.*, 51, 924 (1969).
3. Y. Li, X. Wang, J. Sun, *Chem. Soc. Rev.*, 41, 5998–6009 (2012).
4. B. Neu, A. Voigt, R. Mitlöhner, S. Leporatti, C.Y. Gao, E. Donath, H. Kiesewetter, H. Möhwald, H.J. Meiselman, H. J. Bäumler, *J. Microencaps*, 18, 385–395 (2001).

## 7.2 THE IDEAL FLEXIBLE CHAIN

*Staudinger (1922) demonstrates that a polymer is a covalent chain of N monomers of size a, but he thinks that the configuration is the one of a stick, with a distance between ends $\vec{R} = N\vec{a}$. Kuhn (1940) understands that the chain forms a coil because of the free rotation around the C–C links.*

We focus here on the conformation of an ideal single chain. This corresponds to the case of a chain in a polymer melt or in solution at a temperature $T = \theta$, for which interactions between monomers cancel each other out.

The configuration of an ideal chain is represented by a random walk (RW) on a network of mesh size equal to the monomer length $a$. Each step to a neighboring site has the same probability, $1/d$ ($d = 2$ for a plane network, $d = 3$ for a three-dimensional cubic network). The chain is called a ghost because it can cross its trajectory.

### 7.2.1 End-to-End Distance

The end-to-end distance of an ideal polymer chain is defined by: $\vec{R} = \sum_{i=1}^{N} \vec{a}_i$ (Figure 7.7).

Since the trajectory corresponds to a random path, $\langle \vec{a}_i \rangle = 0$ and $\langle \vec{R} \rangle = 0$. The average value of $\vec{R}$ is zero. Its mean square value is defined by:

$$\langle \vec{R}^2 \rangle = \sum_{i,j=1}^{N} \langle \vec{a}_i \vec{a}_j \rangle = Na^2 \text{ because } \langle \vec{a}_i \vec{a}_j \rangle = a^2 \delta_{ij}.$$

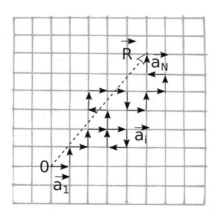

FIGURE 7.7   Random configuration of an ideal chain.

The size of the coil is therefore $R_0 = \left( \langle \vec{R}^2 \rangle \right)^{1/2} = N^{1/2} a$.

## 7.2.2 Gaussian Coil; Entropy Reservoir

The probability $P(\vec{R})$ that the chain starts from zero and reaches the point $\vec{R}$ is given by $P_N(\vec{R}) = \dfrac{\Gamma_N(\vec{R})}{(2d)^N}$ where $\Gamma_N(\vec{R})$ is the number of paths from zero to $\vec{R}$; $(2d)^N$ represents the sum of all paths.

As $\vec{R}$ is the sum of a large number of independent random variables, we can apply the central limit theorem, and we find a Gaussian probability of width $R_0$:

$$P(\vec{R}) = \left( \frac{3}{2\pi Na^2} \right)^{3/2} \exp -\left( \frac{3\vec{R}^2}{2Na^2} \right)$$

Entropy $S(\vec{R})$ is given by Boltzmann's formula:

$$S(\vec{R}) = k_B \, Ln \Gamma_N(\vec{R}) = S_0 - \frac{3}{2} k_B \frac{\vec{R}^2}{R_0^2}$$

A polymer chain is a reservoir of entropy. When the chain is fully stretched, its entropy is zero. For $R = Na$, $\Gamma_N(Na) = 1$, $S(Na) = 0$ leads to $S_0 \approx k_B N$.

If we start from a stretched configuration, $R = Na$, and let the chain relax to $R = 0$, the entropy increases by $\Delta S \approx k_b N$ and the chain takes a quantity of heat $\Delta Q \approx k_B TN$ to the tank that cools down.

A rubber is a set of bridged ideal chains. When it is stretched abruptly, $\Delta S < 0$, which means that the system releases heat that will increase the temperature of the material. On the other hand, when the rubber is released, it cools down. It thus behaves like a thermal machine.

## 7.2.3 Entropic Spring

### 7.2.3.1 Free Energy Argument

For a chain of monomers without interaction, the free energy is:

$$F_{CH} = U - TS = F_0 + \frac{3}{2} k_B T \frac{\vec{R}^2}{R_0^2}$$

When a chain is pulled at both ends with a force $\vec{f}$ (Figure 7.8), the work is $W = \vec{f} \, d\vec{R}$. At $\vec{f}$ and $T$ constant, we minimize $G_{CH}(\vec{R}) = F_{CH} - \vec{f} \vec{R}$ and find:

$$\vec{f} = \left( \frac{\partial F_{CH}}{\partial \vec{R}} \right) = 3 k_B T \frac{\vec{R}}{R_0^2} \text{ or } \vec{R} = \frac{R_0^2 \vec{f}}{3 k_B T}$$

The chain behaves like a spring of stiffness $k = k_B T / R_0^2$. Since $k$ is proportional to $T$, the chain retracts when the temperature is increased. This is the reason why a rubber retracts when heated.

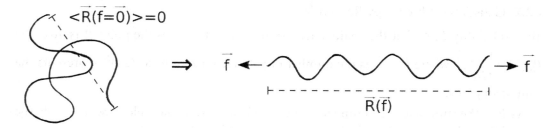

FIGURE 7.8   Ideal chain under traction.

### 7.2.3.2 Scaling Law Argument

The elongation $\vec{R}$ is a length. We have a characteristic length $R_0$. With $R_0$, $f$, and $k_B T$, we form an adimensional variable $u = fR_0/k_B T$. Using a dimensional argument, we can write $R = R_0\, g(u)$.

In the linear response regime, the elongation is proportional to the force, i.e. $R \sim f$. If we assume that $g(u) \sim u^p$, then $R = N^{1/2}\left(R_0 f/k_B T\right)^p \sim f$, which sets $p = 1$. We find: $\vec{R} = \left(R_0^2/k_B T\right)\vec{f}$ i.e. the dependence of $R$ on $N$, $f$, and $T$ obtained by the exact calculation, but the numerical coefficient is lost.

### 7.2.4 Deviations from Ideal Behavior

### 7.2.4.1 Semi-Rigid Chain

*Stiffness effects*: interactions between close neighbors $\langle \vec{a}_i \cdot \vec{a}_j \rangle \neq 0$ for $i, j$.

Example: interaction between first neighbors, valence bond angle (Figure 7.9):

If $V(\theta)$ is the interaction energy, we have:

$$\langle \cos\theta \rangle = \frac{\displaystyle\int \exp\left(\frac{-V(\theta)}{k_B T}\right)\cos\theta \sin\theta\, d\theta}{\displaystyle\int \exp\left(\frac{-V(\theta)}{k_B T}\right)\sin\theta\, d\theta}$$

The angle of the valence bond $-CH_2-CH_2$ is given by $\cos\theta = 1/3$.

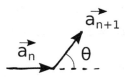

FIGURE 7.9   Definition of the valence bond angle $\theta$.

If $\vec{a}_0$ is set, what is the value of $\vec{a}_n$?

$$\vec{a}_n = \langle \cos\theta \rangle \cdot \vec{a}_{n-1} = \langle \cos\theta \rangle^2 \cdot \vec{a}_{n-2} = \langle \cos\theta \rangle^n \cdot \vec{a}_0$$

*Persistence length* $l_p$: it is defined by: $l_p = \dfrac{1}{a} \sum\limits_{i=0}^{\infty} \langle \vec{a}_0 \cdot \vec{a}_j \rangle$ for an infinite chain, but the series converges quickly.

For interactions between first neighbors:

$$l_p = a \sum_{i=1}^{\infty} \langle \cos \theta \rangle^n = \frac{a}{1 - \langle \cos \theta \rangle}$$

For polyethylene, $l_p \sim 3$ Å. For nitrocellulose, $l_p = 22$ Å. Biological macromolecules are often semi-rigid: $l_p \sim 500$ Å for DNA, $l_p \sim 7$ μm for actin.

*Configuration of a semi-rigid chain*

The chain is a Gaussian coil composed of $Na/l_p$ units of size $l_p$.

$$R_0 = \left( \frac{Na}{l_p} \right)^{1/2} l_p = N^{1/2} (al_p)^{1/2}$$

Short-range interactions between close neighbors do not alter the statistical properties of Gaussian chains.

### 7.2.4.2 Excluded Volume Effect

An ideal chain is called a ghost chain, because monomers can overlap. A chain in good solvent can no longer overlap with itself; as a consequence, it swells, and its statistical properties are changed: the configuration is no longer a Gaussian coil. This case is discussed in Section 7.3.

### 7.2.5 Polymer Weight and Size Measurements

#### 7.2.5.1 Molecular Weights Measurement: Osmometry

Two containers are separated by a semi-permeable membrane (Figure 7.10), which allows the solvent but not the polymer to pass through. The level rises in the compartment containing the polymer chains. The increase in hydrostatic pressure will compensate for the decrease in the chemical potential of the solvent due to osmotic pressure.

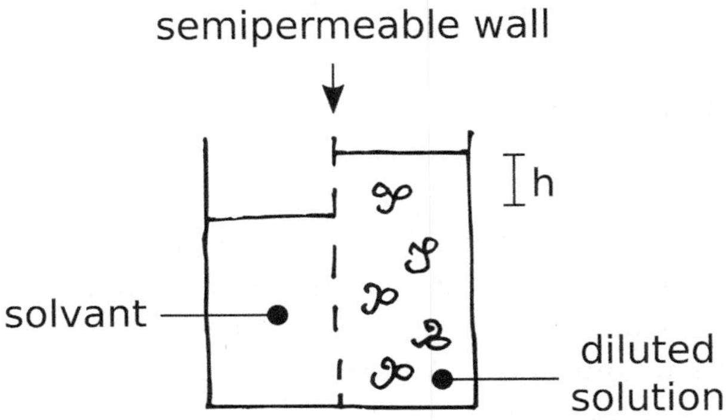

FIGURE 7.10   Osmometry.

The osmotic pressure in dilute solution is given by the law of ideal solutions $\Pi = \dfrac{c}{N}k_BT$, where $c$ is the number of monomers per unit volume. By writing that the decrease of the chemical potential of the solvent is balanced by the hydrostatic pressure corresponding to the rise of the solution level by $h$, we obtain $\Pi = \rho g h$, which yields an easy way to measure $N$.

### 7.2.5.2 Size Measurement: Viscosity

The viscosity of a dilute solution of polymer chains of size $R$ is equivalent to the viscosity of spheres of size $R$ because the chain moves with the solvent it contains (Figure 7.11). Using Einstein's formula for the viscosity of a colloidal solution, we have:

$$\eta = \eta_S\left(1 + \frac{c}{N}R^3 \times k\right),$$

where $\dfrac{c}{N}R^3$ represents the fraction of the volume occupied by the chains, and $k$ is a constant.

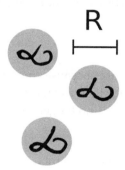

FIGURE 7.11    Viscosity of polymer solutions: the chains behave like balls of size $R$.

The intrinsic viscosity for an ideal chain is written:

$$[\eta] = \frac{\eta - \eta_S}{\eta_S c} = k \times \frac{R^3}{N} \sim N^{1/2}$$

Note: this is the cheapest and most widely used method to measure the size of a polymer chain.

## 7.3 THE SWOLLEN CHAIN

The ideal chain overlaps with itself and describes a RW (Figure 7.12). If the polymer is in a good solvent, each monomer prefers to be surrounded by solvent molecules rather than other monomers, which leads to swelling of the polymer chain. On a network model the chain describes a self-avoiding walk (SAW) (Figure 7.13). Here we will describe the swelling of a flexible polymer chain using the Flory method reformulated by de Gennes. The swelling of the polymer chains controls the rheology of polymer suspensions used in the chemical, cosmetic, and pharmaceutical industries.

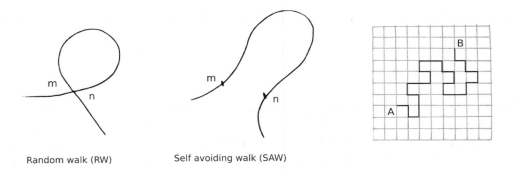

Random walk (RW)          Self avoiding walk (SAW)

FIGURE 7.12   Ideal and swollen chain. The m and n monomers of the ideal chain can overlap, which is not the case for the swollen chain. SAW network model.

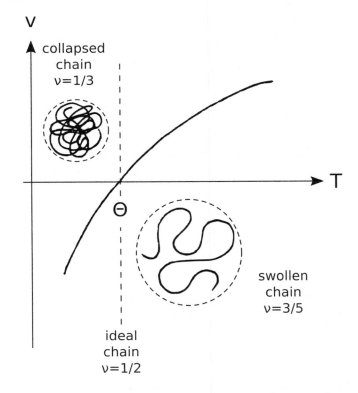

FIGURE 7.13   Variation of the excluded volume parameter $v$ as a function of temperature.

## 7.3.1 Excluded Volume

The excluded volume parameter $v$ describes the interaction between pairs of monomers. It is related to the monomer–monomer interaction potential $V(r)$, which is attractive at long distances (Van der Waals – Section 2.3) and repellent at short distances (hard core). $v$ is given by:

$$v(T) = \int \left( 1 - \exp\left( \frac{V(r)}{k_B T} \right) \right) d^3 r \tag{7.1}$$

The energy of the excluded volume $F_{ve}$ is written as a function of the monomer concentration $c$:

$$F_{VE|volume} = \frac{1}{2} v c^2 \, k_B T \tag{7.2}$$

Dimensionally, $v$ is a volume. In a good solvent, $v = a^3$, where $a$ is the size of a monomer. The excluded volume $v$ cancels out at the temperature $T = \theta$ (Figure 7.13): the polymer chain adopts an ideal chain behavior. For $T < \theta$, $v$ becomes negative: the chain is in bad solvent and collapses into a globule of monomers that expels the solvent. The volume of the polymer chain, proportional to $R^3$ (where $R$ is the radius) is in the order of $Na^3$ with $N$ being the number of monomers. It leads to $R_0 \sim N^{1/3} a$.

### 7.3.2 Flory Calculation [1]

To calculate the repulsion effects between monomers, the chain is considered as a droplet of monomers of size $R$ and concentration $c = \dfrac{N}{R^3}$. The energy of the $F_{ch}$ chain is written:

$$\frac{F_{ch}}{k_B T} = \frac{1}{2} \int v c^2 \, d^3 r + \frac{3}{2} \left( \frac{\vec{R}^2}{R_0^2} \right) \tag{7.3}$$

$F_{ch}$ is the sum of the excluded volume term (obtained by integrating Equation 2 over the volume of the drop) and the entropic term (Section 7.2) which balances swelling (Figure 7.14).

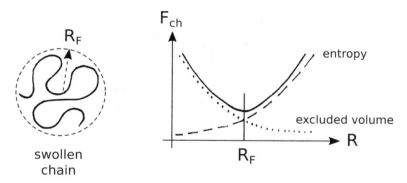

FIGURE 7.14  Swollen chain and energy $F_{ch}$ as a function of size $R$.

By omitting numerical factors and taking $v = a^3$, we obtain:

$$\frac{F_{ch}}{k_B T} \approx N v c + \left( \frac{\vec{R}^2}{R_0^2} \right) \approx v \frac{N^2}{R^3} + \frac{R^2}{Na^2} \tag{7.4}$$

Figure 7.14 shows the shape of the curve $F_{ch}$. The excluded volume term counteracts collapse for small $R$ values, and the entropic term restricts swelling for large $R$ values. $F_{ch}(R)$ has a minimum that gives the size of the chain $R_F$.

$\dfrac{\partial F_{ch}}{\partial R} = 0$ leads to:

$$R = R_F = N^{3/5} a \qquad (7.5)$$

### 7.3.3 Generalization in Dimension $d$

The calculation is easily generalized in a space of any dimension $d$. The excluded volume parameter is $v = a^d$ and the volume $R^d$. The energy $F_{ch}$ becomes:

$$\frac{F_{ch}}{k_B T} \approx a^d \left( \frac{N^2}{R^d} \right) + \frac{d}{2} \frac{R^2}{N a^2} \qquad (7.6)$$

By minimizing the energy of the polymer chain ($\dfrac{\partial F_{ch}}{\partial R} = 0$), we find the general expression of $R_F$:

$$R_F = N^v a \qquad (7.7)$$

with the exponent $v = 3/(d+2)$ (Table 7.2).

TABLE 7.2    Excluded Volume Parameter $v$ as a Function of Dimension $d$

| $d$ | 1 | 2 | 3 | 4 |
|-----|---|-----|-----|-----|
| $v$ | 1 | 3/4 | 3/5 | 1/2 |

The conformations of polymer chains can be observed in dimensions $d$ less than 3:

$d = 1$: the chain is confined to one dimension in a nanotube.

$d = 2$: the chain lies on the surface of an immiscible liquid.

$d = 4$: in a four-dimensional space (which makes sense for numerical simulations), the calculation shows that the chain becomes ideal again.

Although the calculation proposed above is approximate, its validity was demonstrated by P.-G. de Gennes.

### 7.3.4 $n = 0$ Theorem [2]

Pierre-Gilles de Gennes demonstrated that the configuration of a polymer chain describing a RW without overlap is fundamentally related to the physics of phase transitions (Chapter 3).

This finding has two major consequences; it leads to:

1) An exact calculation of $v$ ($v = 0.58$ for $d = 3$),

2) The use of scaling laws and the physics of "blobs."

### 7.3.5 Elasticity of a Swollen Chain [3]

To determine the elasticity of a swollen chain, a force $f$ is applied to both ends of the chain. The elongation $R$ is calculated by a scaling law argument (Figure 7.15).

We write:

$$R = R_F \, g(u) \tag{7.8}$$

with $u = f \, R_F / k_B T$ and $g(u) = u^p$.

Two limit cases can be considered:

i) $u \ll 1$ : in a linear regime, $R \sim f$, therefore $p = 1$, and $R = R_F^2 \left( f / k_B T \right)$.

The chain is a spring of stiffness $k_B T / R_F^2$.

ii) $u \gg 1$ : the chain is highly stretched and $R \approx N$.

Inserting Equation 7.7 into Equation 7.8 gives $R = N^{\nu(1+p)} \left( f / k_B T \right)^p$, which sets $\nu(1+p) = 1$ and thus $p = 2/3$.

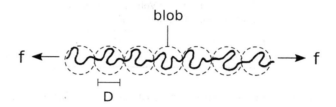

**FIGURE 7.15** Stretched chain: image of "blob."

As a result:

$$R \approx Na \left( f a / k_B T \right)^{2/3} \tag{7.9}$$

*Image of blob* (Figure 7.15): The term "blob" was coined by P.-G. de Gennes and defines a statistical unit, where the chain behaves as an isolated chain containing $g_D$ monomers. When you pull harder and harder on the chain, you straighten it at scales that become smaller and smaller. A "blob" chain portion of size $D$ is defined by: $f D = k_B T$. At scales $r \leq D$, the chain is undisturbed ($D = g_D^{3/5} a$). At scales $r \gg D$, the force aligns the blobs, and we get directly: $R = \dfrac{N}{g_D} D = Na \left( \dfrac{f a}{k_B T} \right)^{2/3}$.

It is remarkable that this law, which shows that the elongation of a chain in good solvent varies as $f^{2/3}$, was calculated in 1974 but only demonstrated experimentally in 2015 using a flexible single stranded DNA.

References

1. P. Flory, *J. Chem. Phys.*, 17, 303–310 (1949).
2. P.-G. De Gennes, *Phys. Lett. A*, 38, 339–340 (1972).
3. P. Pincus, *Macromolecules*, 9, 386–388 (1976).

## 7.4 POLYMER SOLUTIONS

The properties of long chains of flexible polymers in solution are mainly described in good solvent conditions. Polymers in solution have many applications in biology and in industry where they are used as thickening or stabilizing agents. This analysis is extended to gel swelling and polymer mixtures.

### 7.4.1 Three Regimes

Polymer solutions can be characterized by three different regimes, depending on the concentration $c$ in monomers, or the volume fraction $\Phi = ca^3$ (Figure 7.16).

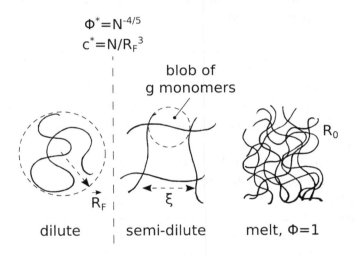

FIGURE 7.16 Schematic representation of the configurations of polymer chains in the different regimes when concentration increases (from left to right).

#### 7.4.1.1 Dilute Solution

In the diluted regime, the chains are separated. The polymer solution is dilute if $\Phi < \Phi^* = N^{-4/5}$.

The chains have a size $R_F = N^{3/5}a$. According to the law of dilute solutions, the osmotic pressure is given by: $\Pi = (c/N)k_B T$ where $c/N$ is the number of chains per unit volume. They come into contact at the concentration $c^* = N/R_F^3$ and entangle for $c > c^*$, corresponding to the semi-dilute regime.

#### 7.4.1.2 Polymer Melt

In the limit of high concentrations, we tend towards a melt.

The polymer melt corresponds to $\Phi = 1$. It is a viscoelastic paste, where the chains are strongly entangled. Flory was the first to show that chains in a melt have the conformation of an ideal chain, $R = R_0$. This hypothesis was validated by neutron scattering experiments, in which a small fraction of chains was labeled with deuterium.

### 7.4.1.3 Semi-Dilute Solution [1,2]

The chains are entangled for $\Phi > \Phi^* = Na^3/R_F^3 \approx N^{-4/5}$. They form a network of mesh size $\xi$ (Figure 7.16). We will derive $\xi$ and the osmotic pressure by using i) the Flory–Huggins theory, which is an average field theory and ii) the scaling laws established by P.-G. de Gennes and based on the analogy with phase transitions.

### 7.4.2 Flory–Huggins Model [1]

#### 7.4.2.1 Polymer–Solvent Solution

Until the discovery of the $n = 0$ theorem, polymer solutions were described by Flory–Huggins' theory.

The monomers and solvent are placed on a lattice. The entropy $\Delta S$ and enthalpy $\Delta H$ of mixing are calculated. The free enthalpy $G = \Delta H - T\Delta S$ per unit volume of solution in a mean field approximation is written as:

$$G = \left(\frac{k_B T}{a^3}\right)\left(\frac{\Phi}{N}Ln\Phi + (1-\Phi)Ln(1-\Phi) + \chi_F\,\Phi(1-\Phi)\right)$$

where the first two terms represent the mixing entropy and the last term is the enthalpy. The parameter $\chi_F$, called the Flory parameter, takes into account the monomer–solvent interactions. The Taylor expansion as a function of $c$ gives the excluded volume term $(1/2)vc^2\,k_B T$, with $v = a^3(1-2\chi_F)$.

The excluded volume cancels out ($v = 0$) for $\chi_F = 1/2$, which defines the temperature $\Theta$. Below the temperature $\Theta$, the chains are in bad solvent, and the solution phase separates, one very concentrated in polymer, the other one very diluted. Above $\Theta$, the polymer is in good solvent, and the solution is homogeneous. In solution in its monomer ($\chi_F = 0$) and $v = a^3$, the polymer is in very good solvent.

The osmotic pressure calculated from $G$ is given by the law of ideal solutions at low concentrations $\Pi = (c/N)k_B T$. At higher concentrations corresponding to the semi-dilute regime, for $\phi > 1/N$, the law $\Pi = (1/2)vc^2\,k_B T \approx c^2$ is obtained (instead of $c^{2.25}$ measured experimentally).

#### 7.4.2.2 Mixture of Two Polymers

If two polymers of polymerization index $N$ and $P$ are mixed, the free enthalpy of mixing in the mean field approximation is written:

$$G\big|_{m^3} = \left(\frac{k_B T}{a^3}\right)\left[\frac{\Phi}{N}Ln\Phi + \frac{(1-\Phi)}{P}Ln(1-\Phi) + \chi_F\,\Phi(1-\Phi)\right]$$

which leads to $v = a^3\left(\dfrac{1}{P} - 2\chi_F\right)$.

If the two polymers are chemically identical, $v = a^3/P$. Excluded volume effects are screened. Thus, if we consider the configuration of a single chain $N$ in a melt of chains $P$, Flory's calculation shows that the chain shrinks when $P$ increases and becomes ideal for $P = \sqrt{N}$.

If the polymers are chemically different, when $\chi_F > 1/2P$, $v$ becomes negative, and the polymers segregate. For Van der Waals monomer–monomer interactions, $\chi_F \approx 1$, and the polymers do not mix. This has considerable industrial consequences because, in order to adapt the properties of plastics, polymer alloys cannot be made in the same way as metal alloys. For this reason, linear or branched copolymers composed of several types of monomers are synthesized. An alternative is the synthesis of diblock copolymers AAAAAAAAABBBBBBBBBBBB which will stabilize the mixtures of polymers A and B, exactly as surfactants allow water and oil to be mixed.

### 7.4.2.3 Swollen Gels

If a gel is placed in a good solvent, Flory–Huggins' free energy with $N$ infinite is written:

$$G\big|_{m^3} = \left(\frac{k_B T}{a^3}\right)\left( (1-\Phi)Ln(1-\Phi) + \chi_F\,\Phi(1-\Phi) \right)$$

This energy is used to calculate the swelling of a gel at equilibrium. If the gel is in good solvent, the equilibrium concentration is the concentration $c^*$.

*Theorem*: A gel formed from cross-linked polymers and immersed in a good solvent swells to reach the concentration $c^*$, corresponding to a volume fraction $\phi^* = N^{-4/5}$, where $N$ is the chemical distance between entanglement nodes.

## 7.4.3 Scaling Laws and "Blob" Model [2]

As soon as the $n=0$ theorem was established, theoretical methods developed for critical phenomena were applied to polymers, in particular scaling laws. We will see that the knowledge of the configuration of an isolated chain allows us to derive the mesh size of the semi-dilute network and the osmotic pressure.

### 7.4.3.1 Mesh Size $\xi$

7.4.3.1.1 Scaling Law Argument  The mesh size $\xi$ is expressed as a function of the length scale $R_F$ and the dimensionless variable $u = \Phi/\Phi^*$. We write $\xi = R_F\, g(u)$, with $g(u) \to 1$ when $u \to 0$ and $g(u) \sim u^p$ for $u \gg 1$.

For $u \gg 1$, $\xi$ becomes independent of $N$ ($\xi_N \sim N^0$), leading to $p = -3/4$, and thus:

$$\xi = a\Phi^{-3/4}. \tag{7.10}$$

We check that $\xi = R_F$ for $\Phi = \Phi^*$ and $\xi = a$ for $\Phi = 1$.

7.4.3.1.2 Blobs Argument  This result can be also derived from the statistical unit of size $\xi$ named "blob" by P.-G. de Gennes (Figure 7.16). At scales smaller than the size of the "blob" $\xi$, the chain behaves like an isolated chain. If $g$ is the number of monomers per blob, the size of the blob is $\xi = g^{3/5}a$. Blobs form a closely packed system, which gives a second relationship $g = c\xi^3$. By eliminating $g$, we find the Equation 7.10. $\xi$ is also the length over which the excluded volume effects are screened, and it provides a measure of the correlation length of the monomer concentration. At scales smaller than $\xi$, the concentration of monomers is very heterogeneous, and at scales larger than $\xi$, it becomes homogeneous.

### 7.4.3.2 Osmotic Pressure $\Pi$

With the same reasoning as above, we can write:

$$\Pi = (c/N)k_B T h(u), \text{ with } h(u)=1 \text{ for } \Phi < \Phi^*$$

and for $\Phi > \Phi^*$, $h(u) \approx u^q$. In the semi-dilute regime, $u \gg 1$, $\Phi$ is dependent on $N$ and $\Pi \approx N^0$, leading to $q = \dfrac{5}{4}$.

As a consequence, for $\Phi > \Phi^*$, $\Pi = k_B T/\xi^3 = (c/g)k_B T \approx (k_B T/a^3)(ca^3)^{2.25}$.

The osmotic pressure $\Pi$ is proportional to the density of chains in the dilute regime and to the density of blobs in the semi-dilute regime. The exponent 2.25 perfectly fits the experimental law $\Pi(c)$.

### 7.4.3.3 Radius R(c)

When the concentration $c$ increases, the chain "deflates," and its size decreases from $R_F$ in the dilute regime to $R_0$ in the melt regime. For the semi-dilute regime, $R(c)$ can be seen as the size of an ideal chain of blobs (Figure 7.17):

$$R(c) = \left(\frac{N}{g}\right)^{1/2} \xi = N^{1/2} a \, \Phi^{-1/8}$$

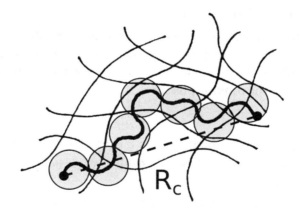

FIGURE 7.17 Schematic representation of $R(c)$.

### 7.4.3.4 Monomer–Monomer Correlation Function g(r)

The correlations between all pairs of monomers are defined by the correlation function $g(r)$:

$$g(r) = 1/c \left[ c(r)\, c(0) - c^2 \right]$$

At short distances $r < \xi$, $g(r) = g_{self}(r) = 1/r^{4/3}\, a^{5/3}$

$g_{self}(r)$ measures the probability of finding a monomer at a distance $r$ from a given monomer, which leads to $g_{self}(r) = n(r)/r^3$, where $n(r)$ is the number $n$ of monomer in a "blob" of size $r$ defined by $r = n^{3/5}\, a$

At large distances, $g(r)$ follows a simple Ornstein–Zernike form (Chapter 3).

$$g(r) = c \frac{\xi}{r} e^{-r/\xi}$$

In terms of a Fourier transform, we obtain:

$$g(q) = c \frac{\xi}{q^2 + \xi^{-2}}$$

$g(q)$ satisfies the compressibility sum rule:

$$g(q = 0) = c\xi^3 = k_B T \frac{\partial c}{\partial \Pi}$$

$g(q)$ is measured by X-ray or neutron scattering. It corresponds to the number of monomers which scatter in phase. In the limit $q\xi \gg 1$, $g(q)$ has to be independent of the concentration $c$:

$$g(q) \sim \frac{1}{(qa)^{5/3}}$$

$g(q)$ is the number of monomers in a blob of size $q^{-1}$.

In summary:

$\xi$ is the mesh size of the transient network of the semi-dilute entangled solution. $\xi$ is also the screening length of the monomer–monomer correlations.

## 7.4.4 Phase Transitions in Polymer Solutions

We describe here the phase transitions in polymer solutions. When the polymer is in solution in a solvent, the separation into polymer-rich and polymer-dilute phases has practical applications for polymer fractionation, with the longest chains segregating first when the temperature is decreased. By precipitating polymer mixtures, bi-continuous phases used in ultrafiltration can be produced.

### 7.4.4.1 Polymer–Solvent Mixture

7.4.4.1.1 Experimental Results    We describe here the case of polystyrene for four molecular weights ($N = 400$ (A), 860 (B), 2,400 (C), 12,000 (D)) in cyclohexane [3]. By lowering the temperature of a solution of concentration $c$ in monomers, we can see by eye that the system separates into two phases: one is turbid, rich in polymer, and the other one is more limpid. By varying the concentration, or equivalently the volume fraction, the phase diagram is built for the different molecular weights (Figure 7.18).

The phase diagram in Figure 7.18 shows that the coexistence curves have a maximum, which is the critical point of demixing, corresponding to a critical temperature $T_c$ and a fraction of monomers $\Phi_c$, which depend on $N$, and that at high concentration, they have a common asymptote. We will calculate the critical temperature.

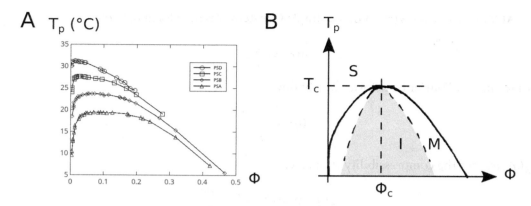

FIGURE 7.18   (A) Precipitation temperature $T_p$ as a function of the volume fraction of PS in cyclohexane. (B) Definition of the coexistence curve, the critical point, and the spinodal curve.

7.4.4.1.2 Modeling   We have seen that the interaction between monomers for chains in solution is described by the excluded volume parameter $v = a^3\left(1 - 2\chi_F\right)$, which cancels out at temperature $\Theta$.

If $v$ is negative, there is a separation in two phases when the temperature is lowered at fixed concentration. The curve $T_p(\Phi)$ shown in Figure 7.18 is the coexistence curve that we will describe.

This transition is modeled from Flory–Huggins' free mixing energy density:

$$G\big|_{m^3} = \left(\frac{k_B T}{a^3}\right)\left(\frac{\Phi}{N}Ln\Phi + (1-\Phi)Ln(1-\Phi) + \chi_F\,\Phi(1-\Phi)\right) = \frac{k_B T}{a^3}g(\Phi)$$

The total enthalpy is $G = \Omega G\big|_{m^3} = nk_B Tg(\Phi)$ where $\Omega$ is the volume of the solution, $n = n_1 + n_2$ is the total number of molecules, $n_1$ the number of solvent molecules, and $n_2$ the number of monomers. We have: $\Phi = n_2/(n_1 + n_2)$ and $\Omega = (n_1 + n_2)a^3$.

7.4.4.1.3 Thermodynamics of Binary Mixtures   Let $\mu_1$ be the chemical potential of the solvent, $\mu_2$ the chemical potential of the monomer, $\mu$ the exchange chemical potential, $\mu = \mu_1 - \mu_2$, and $\Pi$ the osmotic pressure.

Taking $n_1$ and $n_2$, or $\Omega$ and $n_2$ as independent variables, we have:

$$dG = \mu_1 dn_1 + \mu_2 dn_2 = \mu dn_2 - \Pi d\Omega$$

which define the chemical potentials and the osmotic pressure in terms of $g$ and $g' = dg/d\Phi$:

$$\mu_1 = \frac{\partial G}{\partial n_1}\bigg]_{n_2,T} = k_B T\left(g - \Phi g'\right)$$

$$\mu_1 = \frac{\partial G}{\partial n_2}\bigg]_{n_1,T} = k_B T\left(g + (1-\Phi)g'\right)$$

$$\mu = \frac{\partial G}{\partial n_2}\Big]_{\Omega,T} = k_B T g'$$

$$\Pi = -\frac{\partial G}{\partial n_2}\Big]_{n_1,T} = \frac{k_B T}{a^3}\left(-g + \Phi g'\right)$$

7.4.4.1.4 Stability of Solutions: Shape of $G(\Phi)$: The Bitangent Construction    By writing the equality of chemical potentials $\mu_I = \mu_{II}$ in both phases, we show that the system separates into two phases of compositions $\Phi_I$ and $\Phi_{II}$. If $G''$ is positive, the mixture is either stable or metastable and separates by nucleation and growth of polymer-rich droplets (Figure 7.19), and it becomes unstable if $G''$ is negative: concentration fluctuations are amplified. $G'' = 0$ defines the spinodal curve (Figure 7.18B).

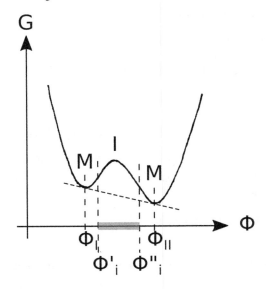

FIGURE 7.19    Shape of $G(\Phi)$ corresponding to a phase separation $(T < T_c)$. If $G'' > 0$, the solution is stable $(\Phi < \Phi_I$ and $\Phi > \Phi_{II})$ or metastable. If $G'' < 0$, the solution is unstable $\Phi_i' < \Phi < \Phi_i''$).

The shape of $\mu(\Phi)$ can be analyzed by analogy with Van der Waals isotherms (equality of surface areas) (Figure 7.20).

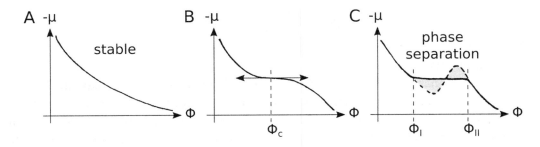

FIGURE 7.20    Curve $\mu(\Phi)$ at different temperatures.

The critical point is defined by:

$$\frac{d\mu}{d\Phi} = \frac{d^2\mu}{d\Phi^2} = 0, \text{ i.e. } g'' = g''' = 0, \text{ which leads to } \Phi_c = \frac{1}{\sqrt{N}} \text{ and } 1 - 2\chi_F = \frac{T_c - \Theta}{\Theta} = -\frac{1}{\sqrt{N}}.$$

The spinodal is defined by: $g'' = 0$, i.e. $2\chi_F = \dfrac{1}{N\Phi} + \dfrac{1}{1-\Phi}$

### 7.4.4.2 Polymer–Polymer Mixture

If two polymers A and B of polymerization indices $N$ and $P$ are mixed, the mixing free enthalpy in the mean field approximation is written:

$$\Delta G\left(\frac{a^3}{k_B T}\right) = \frac{\Phi}{N} Ln\Phi + \frac{(1-\Phi)}{P} Ln(1-\Phi) + \chi_F \Phi(1-\Phi)$$

where $\Phi$ is the fraction in monomers A.

$v = a^3\left(\dfrac{1}{P} - 2\chi_F\right)$ becomes negative for very low values of $\chi_F$.

For the symmetrical case $N = P$, the phase diagram has the shape shown in Figure 7.21.

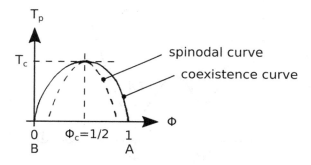

FIGURE 7.21   Spinodal and coexistence curves for a symmetrical polymer–polymer mixture.

The critical point is defined as before. In the symmetrical case, we find $\Phi_c = 1/2$ and $k_B T_c = NU/2$, where $U = \chi_F k_B T$ is the interaction between monomers A and B. The critical temperature is proportional to the polymerization index $N$. It is therefore very high, which explains why there is always a separation between two phases at room temperature.

The spinodal equation is given by:

$$2U = \frac{k_B T}{N}\left(\frac{1}{\Phi} + \frac{1}{1-\Phi}\right).$$

### 7.4.4.3 Summary

In conclusion, phase separation for long polymer chains in solution is characterized by the fact that $\Phi_c \rightarrow 0$, $T_c \rightarrow \Theta$ when $N$ becomes very large. For mixtures of polymer melts, if $U$ is the interaction between monomer A and monomer B, we must compare $NU$ with $k_B T$. Polymers separate because entropy is too low compared to enthalpy to promote mixing.

References
1. P. Flory, *Principles of Polymer Chemistry*. Cornell University Press, 1967.
2. P.-G. De Gennes, *Scaling Concepts in Polymer Physics*. Cornell University Press, 1979.
3. A.R. Shultz, P.J. Flory, *J. Am. Chem. Soc.*, 74, 4760–4767 (1952).

## 7.5 POLYMERS AT INTERFACES

Polymers are used as additives in the formulation of many food products, pharmaceuticals, cosmetics, and paints. They may serve as thickeners and stabilizers. They are also used to modify the wetting properties of substrates and in adhesion, to glue a rubber on a substrate, by interpenetration of long polymer chains with the rubber.

### 7.5.1 Historical Background: From Indian Ink to Polymer Corona

The Indian ink that French people call *encre de Chine* "Chinese Ink" is actually a discovery of the ancient Egyptians circa 4000BC. To disperse particles of carbon black in water, they added Arabic gum to the water: this is a long polysaccharide soluble in water and adsorbed on the carbon surface. In pure water, the colloidal suspension readily flocculates. With Arabic gum, the suspension is stable for years (Figure 7.22).

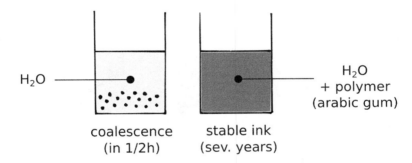

FIGURE 7.22  The Indian ink is stabilized by arabic gum.

We can also mention the stabilization of colloidal gold suspensions by Faraday (1857) [1]: gold colloids are charged and form a red solution in pure water, which precipitates and becomes opalescent and bluish with addition of salt. Faraday stabilized the suspension using albumin, which no longer precipitates with salt. The interpretation came many years later when chemists initiated the steric stabilization of latex for paints (Osmond (ICI) [2]) which was modeled ten years later by Alexander [3], Joanny, Leibler, and de Gennes [4].

Bare carbon grains attract each other by Van der Waals forces (Section 2.3) and stick together with an energy $U \approx \dfrac{k_B T \, R}{d} >> k_B T$, where $R$ is the size of the particles and $d$ their distance in contact (i.e. atomic distance when flocculation occurs) (Figure 7.23A).

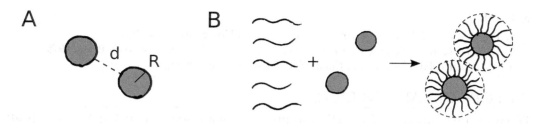

FIGURE 7.23   (A) Flocculation of two "bare" carbon grains. (B) Steric stabilization by the formation of a polymer "corona."

If polymers are grafted onto the grains they form a "corona" that prevents the grains from approaching when in a good solvent (Figure 7.24B).

The polymers at the interfaces can give rise to the three configurations shown in Figure 7.24:

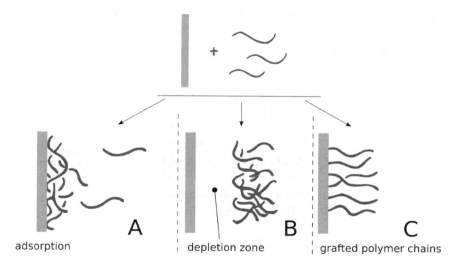

FIGURE 7.24   (A) Adsorption (excess surface area $\Gamma > 0$). (B) Depletion ($\Gamma < 0$). (C) Grafted polymer chains.

Simple *adsorption* of polymers in solution, *depletion*, and *chemical grafting* of long polymer chains, which are discussed below.

### 7.5.2 Adsorbed Polymers: The Self-Similar Grid

We describe here the adsorption of neutral and flexible polymers, in good solvent.

This adsorption plays an important role for numerous applications as illustrated in Figure 7.25:

- *Steric stabilization*:

  When the polymer adsorbs to grains, it forms a corona that will create a steric repulsion between the grains and prevent them from welding under the effect of Van der Waals' attractive interactions.

- *Loss of additives in enhanced oil recovery: Thickening agents*

  If you push oil with water to extract more oil, you will finally collect water because of Saffman Taylor's instability [5]: if a viscous liquid is pushed by a non-viscous liquid, the forehead front between the two liquids is unstable. Water fingers are formed that interdigitate with oil, and finally we collect the water at the exit and not the oil! It is therefore necessary to increase the viscosity of the water by adding polymer. Extremely small quantities of polymers are sufficient to "thicken" the water. Xanthan produced by bacteria or fungi is used. Unfortunately, the polymer adsorbs on most sandstones, and the viscosity drops dramatically. The surface of the rock is chemically very heterogeneous, and it is difficult to find polymers that do not adsorb.

  When the price of oil fell, xanthans were used in the food industry as a thickening agent, in particular for dairy products and jams.

- *Formation of gel by bridging*

  If the macromolecules are of high molecular weight and if the colloidal suspension is sufficiently concentrated, a gel is produced. This particle aggregation can either be a disaster or be very useful. This process is used for water purification. It also makes it possible to extract certain components of petroleum by selective flocculation, a process that is less costly than distillation.

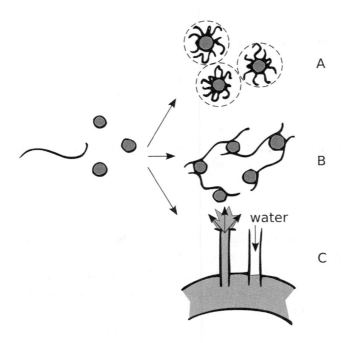

FIGURE 7.25  Adsorption of polymer leading to (A) stabilization or (B) flocculation. (C) Oil recovery.

We will study the adsorption of neutral, long flexible polymers within the high adsorption limit ($U_{solid/solvent} - U_{solid/monomer} > k_B T$).

### 7.5.2.1 Characterization of the Adsorbed Polymer Layer

There are four main experiments that allow the characterization of the adsorbed layer (Figure 7.26):

1) *Surface excess $\Gamma$(monomers/m²)*

   The number of adsorbed monomers per unit area $\Gamma$ is measured by weighing (using ultra-sensitive acoustic methods) or by radioactive tracers. It is found to correspond to a dense monolayer of monomers:

   $$\Gamma \sim a^{-2}, \; a = \text{monomer size}.$$

   Two models can account for this result: a compact monomer layer or a diffuse layer. Sedimentation measurements of polymer-coated particles allow us to distinguish between these two solutions.

2) *Hydrodynamic thickness of polymer-coated particles*

   The mobility $\mu$ is given by the Stokes law:

   $$\mu_{bare} = \frac{1}{6\pi\eta R} \quad \mu_{coated} = \frac{1}{6\pi\eta(R + e_H)}$$

   and the diffusion coefficient by Einstein's relationship:

   $$D_{bare} = \frac{k_B T}{6\pi\eta R} \quad D_{coated} = \frac{k_B T}{6\pi\eta(R + e_H)}$$

   By measurements of sedimentation, inelastic light scattering (ILS), or simply viscosity, one can determine $e_H$. The results are in favor of the diffuse layer model. The experiments performed with polystyrene latex coated with POE in solution in water leads to $e_H = aN^x$, with $x = 0.58$.

3) *Ellipsometry*

   The reflection of polarized light is recorded at the Brewster angle $i = i_B$, for which the induced dipoles are in the direction of the reflected beam. Since light is a transverse excitation, the reflected intensity is zero. A small change in the surface condition will restore light intensity. It is an extremely sensitive zero method to detect surface modifications. From the ellipsometric data, we measure the excess area $\Gamma$ and the first moment $e_e$ of the monomer distribution $\Phi(z)$, defined as:

   $$\Gamma = \frac{1}{a^3} \int \Phi(z)dz \text{ and } e_e = \frac{\int z\Phi(z)dz}{\int \Phi(z)dz} \tag{7.11}$$

Kawaguchi et al. [6] found $e_e = aN^{0.4}$ using the metal (chrome)/PS/CCl$_4$ system.

4) *Detailed profiles $\Phi(z)$*

The detailed profiles have been characterized by "evanescent wave induced fluorescence" (EWIF) and neutron scattering. Since the signal through a wall is too weak in neutron scattering, Loic Auvray studied neutron diffraction by a set of beads on which the polymer is adsorbed.

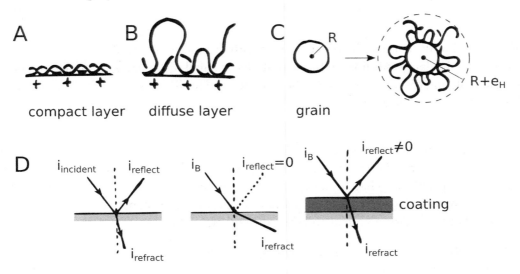

FIGURE 7.26 Characterization of an adsorbed polymer layer (A) by weight: *surface excess* $\Gamma$; (B) by sedimentation: *hydrodynamic length $e_w$*; (C) by ellipsometry: *thickness $e_e$*.

### 7.5.2.2 Description of the Adsorbed Layer: The Self-Similar Grid

The mean-field calculations do not give a good description of the profile $\Phi(z)$, because they assume that the polymer concentration is uniform and cannot include the large fluctuations. We present here the derivation of Pierre-Gilles de Gennes, which leads to the profile with almost no calculations! We show the construction of the fractal self-similar grid (Figure 7.27).

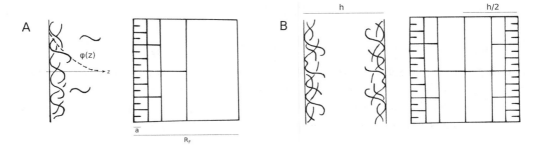

FIGURE 7.27 (A) Description of the adsorbed polymer layer: the self-similar grid $\xi = z$. (B) Repulsion between two adsorbed polymer layers.

We have seen that a semi-dilute solution can be pictured by a compact stacking of blobs of size $\zeta$ containing $g$ monomers. $\zeta = g^\nu a$ (because the blob is a swollen polymer chain) and $g = c\zeta^3$, because the blobs are compact. These two relations led to $\zeta = a(ca^3)^{-3/4} = a\Phi^{-3/4}$.

For the self-similar grid, de Gennes states that there is only one length to describe the polymer layer, which leads to $\zeta = z$, the distance to the wall. $\zeta = z = a\Phi^{-3/4}$ lead to:

$$\Phi(z) = (a/z)^{4/3} \tag{7.12}$$

There is a slow decrease of the concentration with distance to the wall.

The self-similar grid is limited by two cut offs: at small distances, $z_{min} = a$, the monomer size, and at long distances, $z_{max} = R_F$, the polymer size in dilute solution far from the wall. If the solution is semi-dilute,

$$z_{max} = \zeta_{bulk}.$$

Comparison with experimental data: Equation 7.11 with Equation 7.12 for $\Phi(z)$ leads to $\Gamma = a^{-2}$, and $e_e \approx N^{2/5}$, the integral being dominated by $R_F$. The hydrodynamic length $e_H \approx R_F$. The profile $\Phi(z)$ had been measured by neutron scattering by Auvray and Cotton [7] for the system $SiO_2$–PDMS–$C_{12}H_{12}$. There is a perfect agreement between the theoretical model of the self-similar grid and the experimental data.

### 7.5.2.3 Colloid Protection: Repulsion between Two Plates

The plates are left to incubate for a long time to reach the excess surface at equilibrium. The plates are moved closer at constant $\Gamma = \Gamma_{eq}$ as shown Figure 7.27B. The time taken for the chains to re-organize is extremely long, and $\Gamma$ remains constant. There is then a repulsion $U$ between the two plates, caused by interactions of excluded volume between corona.

Between the two plates, monomers give rise to an osmotic pressure $\Pi$. The force $F$ between the two plates vs. the distance $h$ is:

$$F = -dU/dh = \Pi(h/2) \approx k_B T/h^3$$

Leading to:

$$U(h) \approx k_B T/h^2 \tag{7.13}$$

### 7.5.3 Depletion: Flocculation, Size Exclusion Chromatography

When a polymer solution is in contact with a repellent wall, a polymer-free region appears on the wall, called a depletion layer. This phenomenon was first discovered in 1954 by Asakura and Oosaka [8] for dilute solutions and described for semi-dilute solutions using scaling laws for polymers by Leibler, Joanny, and P.-G. de Gennes [4].

### 7.5.3.1 Structure of the Depletion Layer

Figure 7.28 shows the depletion layer:

- For a dilute solution, its thickness is $R_F$.

- For a semi-dilute solution, its thickness is the screening length $\xi$. The polymer volume fraction $\Phi_s$ at the plate can be estimated by the equality of the chemical potential of the solvent, leading to:

$$\Pi = \frac{k_B T}{\xi^3} = \frac{k_B T}{a^3} \Phi_S \text{ leading to : } \Phi_S = \Phi_b^{9/4}$$

The concentration profile is given by a scaling law $\Phi(z) = \Phi_b \left(\dfrac{z}{\xi}\right)^m$, $\Phi(a) = \Phi_S = \Phi_b^{9/4}$ leading to $m = 5/3$.

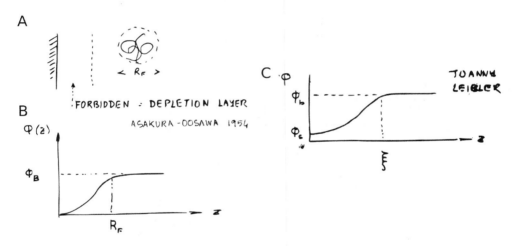

FIGURE 7.28  (A) Schematic representation of a depletion layer. Profile of the monomer density near a repulsive wall (B) for a dilute solution; (C) for a semi-dilute solution. (de Gennes drawings; ©private archives.)

Depletion layers have also been studied at the liquid–air interface. At the free surface of a polymer solution, if the surface tension $\gamma_P$ of the polymer is lower than the surface tension $\gamma_L$ of the solvent, the polymer is adsorbed, and the surface tension $\gamma$ is decreased. On the other hand, if $\gamma_P > \gamma_S$, a depletion layer is formed at the surface, and the surface tension $\gamma$ is increased:

$$\gamma = \gamma_L + \Pi \xi = \gamma_L + k_B T / \xi^2$$

The increase of $\gamma$ corresponds to the work performed against the osmotic pressure $\Pi$ to expel monomers from the depletion layer.

### 7.5.3.2 Interaction between Colloidal Particles via Depletion Layers

Because the osmotic pressure is the driving force to push particles in contact, the attraction between the particle is large in the semi-dilute regime.

Figure 7.29 shows the attraction between two plates and two grains, caused by the osmotic pressure. The energy $F$ vs. the distance $D$ between the plates is expressed as:

$$-\frac{\partial F}{\partial D}\bigg|_{D=0} = \Pi(\Phi_0) = \frac{k_B T}{\xi^3}$$

$$-F_0 = \frac{k_B T}{\xi^2}$$

The depletion adhesion energy per unit area is $W_{adh} = -F_0 = k_B T/\xi^2$.

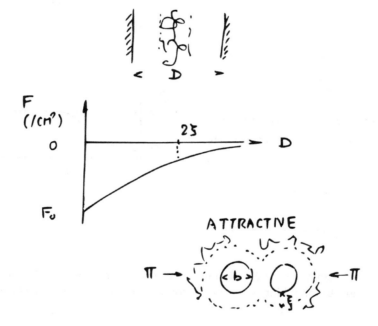

FIGURE 7.29  Attraction between (A) two plates; (B) two grains. (de Gennes drawings; ©private archives.)

Two grains of radius $b$ larger than the correlation length $\xi$ have a contact area $\xi b$. Their total interaction energy is given by:

$$U = -(k_B T/\xi^2) \times \xi b = -k_B T b/\xi$$

### 7.5.3.3 Experimental Measurement of the Depletion-Induced Adhesion Energy

It has been shown that non-adsorbing, water-soluble polymers can induce an attraction of phospholipid bilayers. The adhesion energy $W_{adh}$ induced by the depletion of dextran has been measured experimentally on lipid vesicles and living cells using the dual pipette technique shown Figure 7.30.

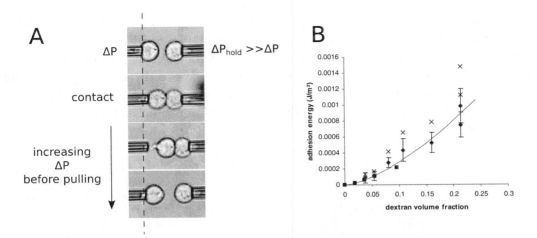

FIGURE 7.30  (A) Dual pipette aspiration techniques to measure the separation force between two non-adherent cells sticking together by depletion attraction; (B) Plot of adhesion energy vs. dextran concentration. The line is the de Gennes theoretical law (Equation 7.14). (Adapted from [9].)

Two cells, held under weak aspiration by micropipettes, were placed in contact and 1 s later became adherent due to depletion of dextran.

The separation process $F_S$ is measured by the dual pipette technique. One cell is held by the right micropipette under strong aspiration. The aspiration applied to the other cell is increased and the right micropipette displaced away. Either the cell leaves the left micropipette or both cells separate (Figure 7.30A). The separation force $Fs$ is deduced from the last aspiration pressures before the cell's detachment and analyzed theoretically. From the measure of $Fs$, one can derive the adhesion energy $W_{adh}$ per unit area by the Johnson–Kendall–Roberts (JKR) [10] formula:

$$W_{adh} = \frac{2Fs}{3\pi R}$$

P.-G. de Gennes has derived the expression of $W_{adh}$ per unit area as a function of the volume fraction $\varphi$ of polymers, which leads in good solvent to:

$$W_{adh} = \left(\frac{k_B T}{a^2}\right)\Phi^{1.5} \tag{7.14}$$

where $k_B T$ is the thermal energy and $a$ the size of a monomer. The theoretical fit $W_{adh}(\Phi)$ is shown in Figure 7.30B.

### 7.5.3.4 Size Exclusion Chromatography (SEC)

SEC is used primarily for the analysis and the separation of large molecules such as proteins or polymers. This process is performed in a column tightly packed with micron-scale beads containing pores of different sizes. These pores may be depressions on the surface or channels through the beads. As the dilute polymer solution travels down the column, small particles enter into the pores, whereas large particles cannot enter into as many pores. Larger molecules therefore flow through the column more quickly than smaller molecules, that is, the smaller the molecule, the longer the retention time. Some supports used in SEC can be described as "fractals" in the size range of macromolecules (10–500 Å) as shown in Figure 7.31. This approach allowed the characterization of a chromatographic column by only two parameters: $D_f$ the fractal dimension of porous grains and $L$ the dimension of larger pores. We show in Figure 7.31 the excluded volume associated with a semi-dilute polymer solution.

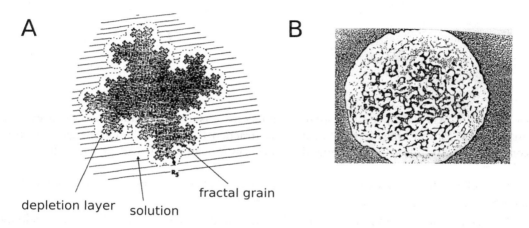

A

B

depletion layer   solution   fractal grain

FIGURE 7.31   The depletion layer is the excluded volume for the polymer coils in a fractal grain. (Extracted from [11].)

### 7.5.4 Chemical Grafting: Polymer Brush

*7.5.4.1 Conformation of Grafted Polymers*

Are polymer chains grafted to the surface of the particles actually stretched like spiky hair?

We consider a grafted surface with long chains of polymerization index $N$. The grafting density is $\Sigma = 1/D^2$, where $D$ is the distance between grafting points. The brush is immersed in a good solvent. There are two regimes.

i)   *"Mushroom" regime*: If $D \gg R_F$, the chains occupy a volume $R_F^3$ where $R_F = N^{3/5}a$ is the size of the free chain (Flory radius). They are not stretched as compared with the situation for a dilute solution (Figure 7.32).

ii)   *"Brush" regime*: If $D \ll R_F$, the chains get stretched and form a brush (Figure 7.33).

If $L$ is the thickness of the brush, the Flory energy per chain can be written as the sum of an excluded volume term and an entropy term (Section 7.3 and Figure 7.34):
$F_{ch}/k_BT \approx N\nu\left(N/LD^2\right)+\left(L^2/R_0^2\right)$ with $\nu$ the excluded volume ($\nu = a^3$ in good solvent) and $R_0$ the radius of the ideal chain. By minimizing $F_{ch}$ with respect to $L$, the equilibrium

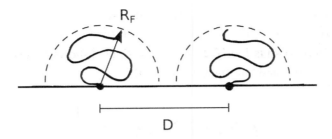

FIGURE 7.32 "Mushroom" regime, $D \gg R_F$.

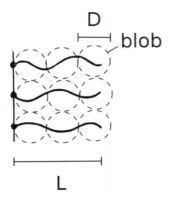

FIGURE 7.33 "Brush" regime, $D \ll R_F$.

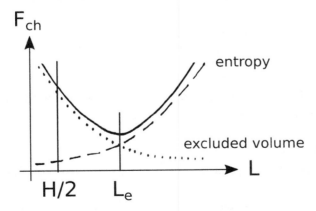

FIGURE 7.34 Energy of the chain as a function of thickness.

thickness of the brush $L_e$ is obtained: $L_e = Na\left(a^2/D^2\right)^{1/3} = Na\Phi_S^{1/3}$ where $\Phi_S$ is the surface fraction of grafted sites ($\Phi_S = a^2/D^2$).

This result can also be readily found by using the blobs argument (Figure 7.33). A chain is described as a string of blobs of size $D$ and containing $g_D$ monomers ($g_D^{3/5} a = D$). We find:

$$L_e = \frac{N}{g_D}D = Na\left(\frac{a^2}{D^2}\right)^{1/3}.$$

7.5.4.1.1 Repulsion between Two Brushes    Two grafted surfaces are approached at a distance $H$ (Figure 7.35). If $H < 2L_e$, the brushes do not interpenetrate but compact themselves. The thickness of each brush is $L = H/2$. The curve $F_{ch}(L)$ shows that the energy per chain is dominated by the excluded volume (Figure 7.34).

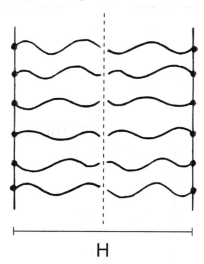

H

FIGURE 7.35    Steric repulsion between two grafted surfaces.

The repulsion energy per unit of surface area in contact is written:

$$U = 2vk_BT\frac{2N^2}{H}\frac{a^3}{D^2} \approx N^2\Phi_S^2\frac{k_BT}{Ha}.$$

The repulsive force: $f = -(\partial U/\partial H) \propto 1/H^2$.

Experimentally, this steric repulsion force can be measured with Israëlashvili's surface force apparatus (Sections 2.3 and 2.5).

## 7.5.5 Concluding Remarks

Grafted or adsorbed polymer brushes are used to stabilize colloidal suspensions and emulsions.

In biotechnology, water-soluble polymer brushes are used to prevent proteins from adsorbing on implants. Another example is that of "stealth" vesicles used as drug carriers. Lipid vesicles injected into the blood are cleared in less than an hour, while the lifespan of vesicles is several days if they are coated with polyethylene glycol (PEG).

The depletion of polymers on non-adhesive substrates is used to precipitate all types of objects, from colloidal particles to living cells.

Finally, polymer brushes can play a role as adhesion promoters. The adhesion energy of a rubber on glass is 50 to 100 times higher if glass is grafted with long polymer chains, which interpenetrate with the rubber.

References

1. M. Faraday, *Philos. Trans. R. Soc. Lond.*, 147, 145–181 (1857).
2. D.W.J. Osmond, D.J. Walbridge, *J. Polym. Sci. Part C.*, 30, 381–391 (1970).
3. S. Alexander, *J. Phys.*, 38, 983–987 (1977).
4. J.F. Joanny, L. Leibler, P.-G. De Gennes, *J. Polym. Sci. Part B.*, 17, 1073–1084 (1979).
5. P. Saffman, G. Taylor, *Proc. R. Soc. A.*, 245, 312–329 (1958).
6. M. Kawaguchi, S. Hattori, A. Takahashi, *Macromolecules*, 20, 178–180 (1987).
7. L. Auvray, J.P. Cotton, *Macromolecules*, 20, 202–207 (1987).
8. S. Asakura, F. Oosaka, *J. Chem. Phys.* 22, 1255–1256 (1954).
9. Y.-S. Chu, S. Dufour, J.P. Thiery, E. Perez, F. Pincet, *Phys. Rev. Lett.*, 94, 028102 (2005).
10. K.L. Johnson, K. Kendall, and A.D. Roberts., *Proc. R. Soc. Lond. Ser. A.*, 324, 301 (1971).
11. M. Lemaire, A. Ghazi, M. Martin, F. Brochard-Wyart, *J. Biol. Chem.*, 106, 814–817 (1989).

## 7.6 DYNAMICS OF POLYMER MELTS: REPTATION

The dynamic properties of flexible polymers in a melt phase are described in the de Gennes model of "Reptation," where long chains behave like snakes that sneak in through the savannah. (Figure 7.36)

This illustrates the style of de Gennes, who compared scientists to savages in the jungle rather than gardeners in an orchard in his Leçon Inaugurale au Collège de France (1971) and uses the same kind of metaphor in science (snakes for polymers) that poetically brings these ideas to life, hard to communicate in a different way as he points out (Figure 7.36).

FIGURE 7.36   Snake motion in a tube.

Long chains are entangled in a melt because they cannot intersect each other; they make all sorts of complicated knots. These knots are essential for their remarkable dynamic properties. Let's take an ultra-viscous paste, e.g. a so-called Silly Putty, which is a melt of silicone polymers (Section. 8.8). After shaping it into a ball, we note that it bounces like a rubber ball if you drop it, but it spreads like a liquid if you let it rest on a table. This system behaves like a solid at short timespans and flows like a liquid at long timespans. Similarly, if two small Silly Putty balls are put in contact for a very short time, they separate like two solid balls. But if you keep the contact for a few seconds, you notice that they have stuck together and merged partially; when you try to separate them, they flow and form a long polymer thread. This self-adhesion phenomenon makes it possible to glue kilometers of polyethylene pipes for gas transport without adhesives.

### 7.6.1 Viscoelasticity

Polymer melts are extremely viscous: they behave like rubbers at short times and flow like normal liquids if slow perturbations are applied. The interpretation has been known for a long time. Polymer chains are entangled: at short times, the nodes between chains do not have time to untie, and the behavior is the same as that of a cross-linked system. In long timespans, the knots are untied by Brownian motion. The time that separates these two regimes is the creep time $T_r$.

The mechanical properties of polymers are described in the classic textbook by J.D. Ferry [1]. A stress $\sigma$ is applied at time $t=0$ to a polymer rod of section $S$ and length $\ell_0$. Its elongation $\ell(t)$ is monitored as a function of time $t$. Figure 7.37 shows the compliance $J(t)$, defined by the deformation-to-stress ratio, with $\varepsilon = (\ell - \ell_0)/\ell_0 = \delta\ell/\ell_0$ the deformation. The curve $J(t)$ allows the definition of three parameters that characterize a viscoelastic fluid: the elastic modulus $E$, the viscosity $\eta$, and the creep time $T_r$ which separates the elastic and viscous regimes, and is named the reptation time in polymer physics.

To model the curve $J(t)$ of a viscoelastic liquid, the simplest model is the Maxwell model (Figure 7.37B). The tension $\sigma$ is transmitted to the two elements, namely the "dashpot" and the "spring," and generates deformations $\varepsilon_2$ and $\varepsilon_1$ respectively.

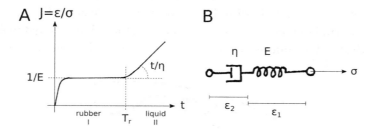

FIGURE 7.37 (A) Representation of compliance $J(t)$ and (B) drawing of the Maxwell model used to describe viscoelastic properties, which consist of a dashpot and a spring in series. $\varepsilon = \varepsilon_1 + \varepsilon_2$.

This leads to the following relations:

$$\sigma = E\varepsilon_1 = \eta\dot{\varepsilon}_2$$

$$\varepsilon = \varepsilon_1 + \varepsilon_2$$

Hence: $(d\sigma/dt) + (E/\eta)\sigma = E\dot{\varepsilon}$

This stress relaxation equation describes all viscoelastic liquids. At short timespans, $J = \varepsilon/\sigma = 1/E$ and at long times, $J = (1/\eta)t = (1/E)t/T_r$ with $T_r = \eta/E$, as shown in Figure 7.36.

### 7.6.1.1 Relaxation Time $T_r$

This relaxation time exists for all systems, and it is generally very small for standard liquids, in the order of $10^{-10}$ s. For polymer melts, $T_r$ is enormous compared to the relaxation time $\tau_0$ of the monomer:

$$T_r = \tau_0 N^a, \tag{7.15}$$

with $3 < a < 3.5$. Taking $\tau_0 = 10^{-10}$ s, one finds $T_r$ in the order of minutes for long chains (of polymerization index $\sim 10^4$).

For short times $t < T_r$ or equivalently, at high frequencies $\omega > 1/T_r$, the behavior of the polymer melt is that of a solid rubber. On the other hand, for long times $t > T_r$ or equivalent at low frequencies $\omega < 1/T_r$, the polymer melt behaves like a liquid.

### 7.6.1.2 Elastic Modulus E

Young's modulus $E$ is large when the distance between points of entanglement is small. $N_e$ is defined as the "chemical distance," i.e. the number of monomers between entanglement nodes. For an elastomer, the elastic modulus is proportional to the density of the crosslinking points:

$$E = \rho \frac{k_B T}{N_e},\tag{7.16}$$

with $\rho = 1/a_0^3$ the density in number of monomers per unit volume, $a_0$ the size of a monomer, and $\rho/N_e$ the density of entanglements.

The entanglement threshold, i.e. the minimum number of monomers necessary to make entanglements, is in the order of $N_e \sim 200$.

### 7.6.1.3 Viscosity η

For a Maxwell type viscoelastic liquid, characterized by a dashpot and a spring in series (Figure 7.29), the relationship between the characteristic Maxwell time (here $T_r$), the elastic modulus $E$, and the viscosity is:

$$\eta = E \cdot T_r.\tag{7.17}$$

*Case $N > N_e$:* from Equations 7.15 and 7.16, we find that:

$$\eta \sim N^a.\tag{7.18}$$

*Case $N < N_e$:* there is no entanglement. To find the corresponding relaxation time, we use Rouse's approach [2], which consists of modeling a polymer chain by a set of balls connected by springs. Intuitively, the diffusion coefficient $D$ of the chain is reduced compared to that of a monomer, $D_0$. It is given by $D = D_0/N$. Therefore, the Rouse relaxation time $\tau_{\text{Rouse}}$, which is the time for the chain to diffuse over its size $R_0$, is defined by $D\tau_{\text{Rouse}} = R_0^2$, i.e. $\tau_{\text{Rouse}} = \tau_0 N^2$, where $\tau_0^{-1} = k_B T/\eta_0 a_0^3$, $\eta_0$ is the monomer viscosity. In addition, the osmotic modulus, of the order $k_B T$ per chain, is $E = \rho(k_B T)/N$.

Using the scaling relationship between viscosity and relaxation time Equation 7.17, leads to:

$$\eta = \eta_0 N\tag{7.19}$$

The viscosity vs. $N$ curve displays two regimes (Figure 7.38). The measure of $\eta$ for different chain lengths provides a simple way to determine $N_e$.

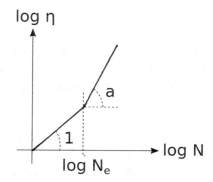

FIGURE 7.38  Evolution of viscosity $\eta$ as a function of $N$ and determination of $N_e$.

## 7.6.2 Reptation Model

This model originates from Sam Edwards' image (1967) [3] which described the chain in a polymer melt as confined in a tube due to topological constraints imposed by other chains. The size of the tube is determined by the distance between entanglements (Figure 7.39).

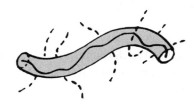

FIGURE 7.39  Edwards' model of a polymer chain confined in a tube. The diameter is the size of a knot to entangle chains, of order 5 nm.

Pierre-Gilles de Gennes added the vision of the Brownian movement of the chains threading through these tubes like a snake in the savannah. This will give rise to the reptation model [4]. P.-G de Gennes explains it to a non-scientific audience by taking a dish of long cooked spaghetti, with one that is colored red. By pulling on the red spaghetti, he shows the complex path due to topological constraints. A polymer melt is equivalent to a spaghetti dish subjected to thermal agitation. The detailed description of such a system seems disproportionately complex: there are more than 60 topological classes of nodes. But we will see that this reptation model is surprisingly simple.

Due to the topological constraints set by other chains, a chain must move in a tube along its axis (Figure 7.40). The lateral dimension $d$ of the tube corresponds to the size of an ideal blob composed of $N_e$ monomers, i.e. $d = N_e^{1/2}a$ and the curvilinear length of the tube is the number of blobs $N/N_e$ multiplied by the size of a blob, i.e. $L = Na/N_e^{1/2}$. The movement is diffusive (Figure 7.40).

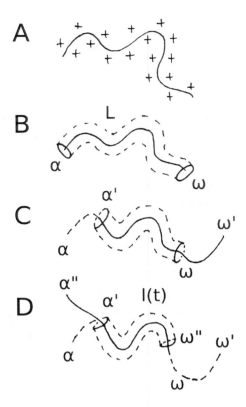

FIGURE 7.40 (A) Chain trapped in a tube formed by the topological constraints (represented by "+") created by the other chains. (B–D) Three successive steps of the Brownian motion of a chain in the tube: (B) at initial time; (C) the chain moved forward from ω to ω′; (D) the chain moved backward from α′ to α″. It loses the memory of its original tube.

### 7.6.2.1 Reptation Time $T_r$

What is the mobility of the chain in its tube? If the confined chain is slid at a velocity $V$, the drag force is the sum of the friction on each monomer, i.e. $F_v = N\eta_0 a_0 V$, which can be written $V = \mu_t F_v$. Here, we define $\mu_t = (\mu_0/N)$ the mobility of the chain in its tube, with $\mu_0$ the mobility of a monomer.

The diffusion coefficient of the chain in its tube is given by the Einstein formula $D_t = \mu_t kT = (D_0/N)$. The time $T_r$ it takes for the chain to diffuse over the length $L$ of the tube, thus losing all memory of its initial configuration, is obtained using the classical diffusion relation $L^2 = 2D_t T_r$, which gives:

$$T_r = \tau_0 \frac{N^3}{N_e}. \tag{7.20}$$

Note that for $N = N_e$, $T_r$ becomes the Rouse time.

### 7.6.2.2 Viscosity of the Melt

The viscosity is the product of the elastic modulus by the reptation time $T_r$, Equation 7.20, i.e.:

$$\eta = \eta_0 \frac{N^3}{N_e^2} \tag{7.21}$$

We notice that there is a continuity for $N = N_e$ with the viscosity of the disentangled melt. Experimentally, an exponent slightly greater than 3 is found.

### 7.6.2.3 Self Diffusion: $D_{self}$

Consider one test chain, labeled to be recognized in the melt. After a time $t$ much larger than the reptation time $T_r$, this chain's center of mass will have moved on a distance $x$, of mean square:

$$x^2 = 2 D_{self} t \tag{7.22}$$

Because the tube is contorted, the self diffusion coefficient $D_{self}$ is different from $D_t$, but it is easily derived from the reptation model. Within the time period $T_r$, the chain has traveled over a curvilinear length $L$, but its center of mass has only moved over a distance equal to its gyration radius $R_0$. So we have two relations: $D_t \cdot T_r = L^2$ and $D_{self} \cdot T_r = R_0^2$, which give $D_{self} = D_t \left( R_0^2 / L^2 \right)$, or equivalently:

$$D_{self} = \frac{D_0 N_e}{N^2}. \tag{7.23}$$

By taking $D_0 \sim 10^{-5}$ cm$^2$/s, $N_e \sim 100$ and $N \sim 10^4$, we get $D_{self} \sim 10^{-11}$ cm$^2$/s.

## 7.6.3 Experiments on Self-Diffusion

The self-diffusivity is indeed a very slow process. The time $t$ required to perform one experiment is related to the distance $l$ over which the diffusion will take place, $D_{self} t \approx l^2$. In usual tracer techniques, $l > 1$ mm and $t \gg 10^9$ sec. To perform such experiments, one must either wait a very long time or try to decrease the diffusion length by several orders of magnitude.

In the first experimental approaches, Klein and Briscoe [5] used deuterated polyethylene as labeled chains diffusing in a protonated polyethylene melt and measured the local concentration of labeled monomers through the amplitude of the IR absorption of the C–D bond. They then had access to the total concentration profile, with a spatial resolution of about 0.1 mm, and the times involved were roughly one month! They measured $D_{self}$ as a function of molecular weight. They got a result very close to the reptation prediction.

A clever technique, namely forced Rayleigh scattering (FRS), was introduced in polymer physics by the group of the Collège de France (F. Rondelez and L. Léger): it allowed them to go down to scales $l \sim 1$ µm and thus to reduce the experimental times to a few minutes [6]. With this technique, the labeled chain carries a photochromic group; this is a cyclic molecule that opens up its cycle upon irradiation, thereby changing its absorption. The sample is illuminated with an optical fringe pattern, which generates bright fringes (where molecules are photo-activated) alternating with dark fringes, leading to a spatially periodic distribution of photo-exited molecules, as shown in Figure 7.41. This periodic distribution of photo-excited molecules acts as an absorption grating for a second laser beam, sensitive to the photo-excited species. After one cuts off the illumination, the grating fades out by diffusion of the labeled polymers. This fading out is monitored with the second laser beam, which measures as a function of time the diffraction pattern due to the grating. The decrease of the diffracted intensity directly reflects the kinetics of the diffusion of the chains. FRS is a tracer technique in which the diffusion length can be reduced down to a few microns, without any loss of accuracy.

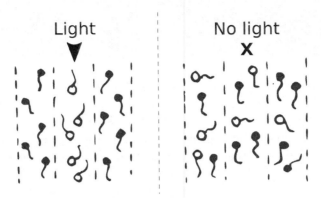

FIGURE 7.41   Forced Rayleigh scattering: a polymer melt (or a solution) containing a few chains labeled with a photochromic dye is exposed to interference fringes of intense light; in the illuminated region, the dye becomes dark and remains transparent elsewhere. The dark and transparent slabs fade out by diffusion. This is monitored by Bragg reflection of light from a low-power laser.

The characteristic spatial scale is the inter-fringe distance or the inverse of the wave number $q$ (with $q = 2\pi / i$, where $i$ is the inter-fringe distance), therefore $D_{self} q^2 = 1/\tau$, with $\tau$ the characteristic diffusion time. The diffracted intensity is given by $I_q \sim I_0 e^{-t/2\tau}$, and so the determination of $\tau$ and $D_{self}$ becomes straightforward.

FRS had been used to measure self-diffusion in polymer solutions. Self-diffusion in entangled polymer solutions seems more difficult a priori than in a melt, as concentration effects have to be taken into account. The chains still cannot cross each other, but the size of the tube is now related to the mesh size $\xi$ of the polymer solution. The chain is a succession of blobs of size $\xi$. We derive $D_{self}$ in the next Section 7.7 using the reptation model applied to semi-dilute polymer solutions.

In the case of polystyrene dissolved in benzene, the group of the Collège de France showed that $D_{self} \approx c^{-1.7 \pm 0.1} N^{-2 \pm 0.1}$, which validates the theoretical law (Section 7.7).

### 7.6.4 Visualization of the Reptation Using Giant Polymers

Following the above-mentioned pioneer works, numerous other experimental studies allowed the validation of the theoretical predictions from the measurements of polymer bulk properties. However, the direct visualization of the reptation motion of a polymer chain was only performed in 1994 by Steve Chu's group [7] using DNA. In the fluorescence microscopy photographs shown in Figure 7.42, a labeled DNA chain is seen in a concentrated solution of identical unlabeled (and therefore invisible) molecules relaxing after being abruptly stretched with an optical trap (via a bead attached to one end).

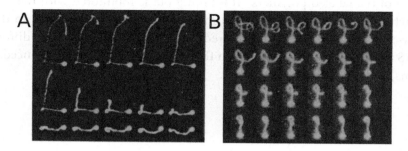

FIGURE 7.42    Reptation of a fluorescent DNA chain within unlabeled chains. (Extracted from [7].)

### 7.6.5 Separation of Charged and Neutral Polymer Chains

If polyelectrolytes can be separated by ultracentrifugation, they move in an electrical field at the same velocity whatever their length: the electrophoretic motility of a charged polymer is independent of the molecular weight of the polymer.

The basic separation principle of polyelectrolytes like DNA, consisting of circulating the polymers through obstacles or in a gel, is based on the model of reptation.

The first DNA separation technique used agarose or polyacrylamide gels electrophoresis. The charged molecules are incorporated in a water swollen gel, where they drift under an electrical field $E$ with a velocity $\mu E$. The electrophoretic mobility $\mu$ is found to decrease with the molecular weight, and this allows for the separation. However, to be in this linear regime, i.e. where velocity is proportional to the electric field $E$, it is necessary to use weak fields ($E \sim 1$ to $300$ V/cm). For a strong electrical field, DNA is stretched, the size of the chain varies as $N$ instead of $N^{1/2}$, and selectivity is lost.

More recently, the advantages of microfluidics and microfabrication have been exploited to produce microchannels containing magnetic particles that form columns or pillars when a magnetic field is applied (Figure 7.43) [8].

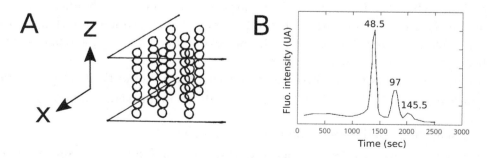

FIGURE 7.43    (A) Schematic drawing of a microfluidic channel part containing columns of magnetic beads that serve as obstacles. (B) Fluorescence detection of the separation of three types of DNA, respectively 48.5, 97, and 145.5 kbp in length ($\times 10^3$ base pairs). (Adapted from [8].)

To determine the dependence on $N$ of the electrophoretic mobility, defined by $\vec{V} = \mu_E(N)\vec{E}$, let us return to our reptation model (Figure 7.44).

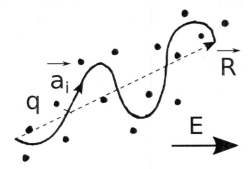

FIGURE 7.44 Drawing of the reptation of a DNA chain across obstacles $q$ is the charge per monomer and $\vec{f_i} = q\vec{E}$ is the force.

The force pulling on the chain in its tube is: $f_t = \sum_i \vec{f_i} \cdot (\vec{a_i}/a) = \rho\vec{E} \cdot \vec{R}$ with the notations of the drawing in Figure 7.37. However, we have seen that the curvilinear velocity is $V_t = (\mu_0/N)f_t$. The velocity of the center of mass is therefore $\vec{V} = V_t(\vec{R}/L)$, which defines electrophoretic mobility: $\mu_E(N) = (\mu_0/N)q(R^2/L) \sim 1/N$. This shows that there is indeed a good selectivity in size.

References

1. J.D. Ferry, *Viscoelastic Properties of Polymers*. New York, NY: Wiley, 1970.
2. P.-G. De Gennes, *Scaling Concepts in Polymer Physics*. Cornell University Press, 1979.
3. S.F. Edwards, *Proc. Phys. Soc.*, 92(1), 9–16 (1967).
4. P.-G. De Gennes, *J. Chem. Phys.*, 55, 572 (1971).
5. J. Klein, B. Briscoe, *Proc. R. Soc. Lond. A*, 365, 53 (1979).
6. L. Léger, H. Hervet, F. Rondelez, *Macromolecules*, 14, 1732–1738 (1981).
7. T.T. Perkins, D.E. Smith, S. Chu, *Science*, 264, 819–822 (1994).
8. P.S. Doyle, J. Bibette, A. Bancaud, J.L. Viovy, *Science*, 295, 2237 (2002).

## 7.7 DYNAMICS OF POLYMER SOLUTIONS

We describe the dynamics of polymer coils in solution in a good solvent. From the laws in the dilute limit, one can extend the dynamical behavior to the semi-dilute limit, using *dynamical scaling laws*. From the dynamics of the melt, we can discuss the self-diffusion in the semi-dilute limit, where the chains are entangled. Dynamical scaling is powerful, because there is only one single length $\xi$, which is the *static* screening length for the excluded volume interactions and also the *dynamic* screening length of the hydrodynamic interactions.

It is important to notice that for binary mixtures, polymer and solvent, we can define three diffusion coefficients: $D_{coop}$, the *cooperative diffusion* coefficient, associated with the fluctuations of the global concentration "$c$," the *self-diffusion* coefficient $D_{self}$ associated with the Brownian motion of the long polymer in the transient polymer network at uniform concentration, and the auto-diffusion coefficient of the solvent $D_0$, which is a constant. Cooperative diffusivity $D_{coop}$ is an increasing function of concentration, because at high osmotic pressure, the restoring force associated with the concentration fluctuations are stronger. On the other hand, $D_{self}$ is a decreasing function of concentration because the solution becomes more entangled.

The analogy with critical phenomena, or the "$n = 0$" theorem, allows the description of the dynamics of a polymer solution using scaling laws, characterized by a dynamical exponent $z$. The theoretical value $z = 3$ is in agreement with experimental data.

### 7.7.1 Single Chain Dynamics

Under the action of an external force $F$ (gravity, electric field, etc.), a linear and flexible polymer chain (of $N$ monomers) moves in a solvent (viscosity $\eta$) with a velocity $V = s_{ch}F$, where $s_{ch}$ is the chain motility. We will describe the mobility coefficient for Rouse and Zimm models. We will discuss also the particular case of a charged chain.

#### 7.7.1.1 Immobile Solvent: Rouse Model [1]

First, let's define the mobility of a monomer, $s_0$. Under the action of a force $f$, the monomer moves at the velocity $V$ given by Stokes–Einstein's law: $f = 6\pi\eta aV$, where $a$ is the size of a monomer. By definition of mobility $V = s_0 f$, we obtain $s_0 = (1/6\pi\eta a)$.

Considering a chain of $N$ monomers, the drag force $F$ is the sum of the friction forces exerted on each monomer, i.e. $F = N6\pi\eta aV$ (Figure 7.45).

We thus find that $s_{ch} = (s_0/N)$. The chain, due to thermal agitation, has a Brownian motion characterized by a diffusion coefficient $D_{ch}$. By using Einstein's relation, we have:

$$D_{ch} = s_{ch}k_BT = \frac{s_0}{N}k_BT = \frac{D_0}{N} \tag{7.24}$$

With $D_0$ the diffusion coefficient of the monomer. Finally, we can define Rouse's relaxation time, $\tau_R$, from the diffusion equation: $D_{ch}\tau_R = R^2$, i.e. $\tau_R = \tau_0 N^2$ for an ideal chain (with $\tau_0$ the characteristic molecular time ~$10^{-10}$ s).

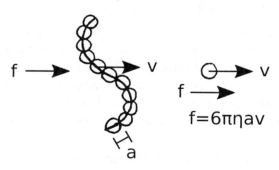

FIGURE 7.45 Drawing of a polymer chain in Rouse model. Each monomer is assimilated to a bead. The polymer is a string of beads moving at velocity $V$ under force $F$.

This situation corresponds to the case of a chain in a polymer melt or in a concentrated solution. In a dilute solution, the model is not valid because the chain carries the solvent molecules with it.

For a fluctuation of monomers at wave vectors $qR > 1$ the dynamical scaling laws assume that $\omega(q)$ is a homogeneous function of $q$ and the characteristic length $R$:

$$\omega(q) = \frac{1}{\tau_0 N^2} \Omega(qR)$$

$\Omega(x) \to 1$ for $x \to 0$, and $\Omega(x) \to x^z$ for $x \to \infty$.

The exponent $z$ is derived from the condition that at large $q$, the dynamics are independent of $R$. It leads to $z = 4$ if the chain is ideal, $R = R_0 = N^{1/2}a$.

### 7.7.1.2 Dragged Solvent: Zimm Model [2]

As shown in the drawing in Figure 7.46, the force $f_n$ applied to the monomer $n$ creates a so-called long range "backflow" velocity field:

$$\vec{V}(r) = \sum_n \vec{f}_n \cdot \frac{1}{6\pi\eta |r - r_n|}. \tag{7.25}$$

In summing the action of all monomers, we find that the chain moves by driving the solvent like a ball of hydrodynamic size $R_H$ comparable to the Flory radius (Section 7.3): $R_H \approx R_F \approx N^\nu$ with $\nu$ the Flory exponent ($\nu = 3/5$ in good solvent and $\nu = 1/2$ in $\Theta$ solvent). Finally, we find: $V = (F/6\pi\eta R_H)$, $s_{ch} = (s_0/N^\nu)$ and the Einstein equation leads to

$$D_{ch} = \frac{k_B T}{6\pi\eta R_H} \sim N^{-\nu} \tag{7.26}$$

The characteristic Zimm time, $\tau_Z$, is given by: $D_{ch}\tau_Z = R_F^2$, i.e. $\tau_Z = (\eta_0 R_F^3/k_B T) \approx N^{3\nu}$.

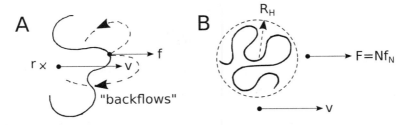

FIGURE 7.46 (A) Backflow induced by the force $f_n$ acting on monomer $n$. (B) Hydrodynamically, a coil behaves like a sphere of radius $R_H \sim R_F$.

In a dilute solution, the polymer chains move by dragging the solvent and behave as spheres of radius $R_H = R_F$. For this reason, the viscosity of a dilute polymer solution is well accounted for by Einstein's formula for a suspension of beads of size $R_F$.

For a fluctuation of monomers at wave vectors $qR_F > 1$ the dynamical scaling laws assume that:

$$\omega(q) = \frac{1}{\tau_z} \Omega(qR)$$

With $\Omega(x) \to 1$ for $x \to 0$ and $\Omega(x) \to x^z$ for $x \to \infty$. The exponent $z$ is derived from the condition that at large $q$, the dynamics are independent of $R_F$. It leads to $z = 3$. This result had been confirmed by inelastic light scattering by M. Adam [4] using extremely long PS molecules ($M = 3,107$ g/mol), in the limit of large $q$ ($qR \approx 10$).

### 7.7.1.3 Backflow Free Case: Debye Screening

We consider a charged chain in solution in motion under an electrical field $\vec{E}$ (Figure 7.47). Screening of the charges along the chain occurs over the Debye length $\kappa_D^{-1}$ (Section 2.4), which can vary typically from 10 to 100 Å, depending on the salt concentration of the solution. According to Debye's theorem, screening of hydrodynamic interactions is identical to screening of electrostatic interactions. Indeed, if we consider a blob of size $\kappa_D^{-1}$, the sum of the electrical forces acting on the monomers and counter-ions cancels out, and according to Equation 7.24, there is no induced flow at long range.

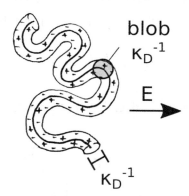

FIGURE 7.47 Drawing of a polyelectrolyte chain in solution under electric field $E$: the Debye screening length $\kappa_D^{-1}$ is also the hydrodynamic screening length.

The charged polymer therefore behaves like a Rouse chain with a monomer mobility $s_0$. Since the electrical force is generically defined as $F = QE$ (with $E$ the electric field and $Q$ the total charge), we obtain: $V = (s_0/N)QE = s_0qE = \mu_E E$ where $q = (Q/N)$ is the monomer charge and $\mu_E$ is the electrophoretic mobility, independent of $N$ ($\mu_E \propto N^0$). Polyelectrolyte chains can thus be only separated by gel electrophoresis. In sedimentation, the chain behaves like a sphere of radius $R_F$.

### 7.7.2 Dynamics of Fluctuations of Polymer Solution: The Cooperative Diffusion Coefficient

We show in Figure 7.47 an illustration of the main characteristic parameters describing the dynamics of polymer solution from the dilute to the semi-dilute regime.

The dynamics of polymer solution are dependent on the concentration and governed by the cooperative diffusion coefficient, $D_{coop}$, as summarized in Figure 7.48 and explained in the sub-sections below.

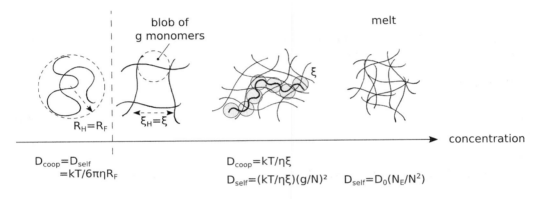

FIGURE 7.48 Dynamics of polymer solutions. In the dilute regime, $D_{coop} = D_{self} = k_B T/6\pi\eta\, R_H$. In the semi-dilute regime, $D_{coop} = D_{blob} = k_B T/6\pi\eta\xi$, and $D_{self} = D_{blob}(g/N)^2$.

#### 7.7.2.1 Sedimentation Coefficient s: Hydrodynamic Screening Length ξ

Each monomer in the solution is submitted to a force $f_m$ ($f_m$ can be a centrifugal force). It leads to a drift velocity $V_P$ of the polymer.

$$V_P = s\, f_m, \tag{7.27}$$

which defines $s$, the sedimentation coefficient.

In the approach of Kirkood and Risemann (see [3] for the detail of the calculations), a monomer M feels the flow field from the neighboring monomers M′ as shown in Figure 7.49. The velocity $V(M)$ of M is the sum of backflow contributions from all sources of forces given by the Oseen tensor (Equation 7.25). The probability of finding a monomer at M and M′ is given by the correlation function $1/c < c(0)c(r) >$. To maintain the solvent in the steady state, a pressure gradient must exactly balance the force per unit volume $cf_m$. Adding this term leads to a probability $1/c < c(0)c(r) > -c = g(r)$.

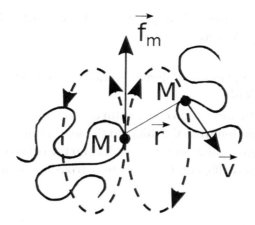

FIGURE 7.49 A monomer M drift in the backflow (indicated by the arrows) induced by another monomer M′ subjected to a force $\mathbf{f_m}$.

The sum leads to:

$$s = \int \frac{1}{6\pi\eta r} g(r)\, dr \tag{7.28}$$

with $g(r) = c\left(\xi/r\right)e^{-r/\xi}$

This formula shows that the backflows are screened, which leads to the fundamental result:

*The correlation length $\xi$ is the screening length of the hydrodynamic interactions.*

The extension of $s$ derived from Equation 7.27 is:

$$s = \frac{g}{6\pi\eta\xi} \sim c\xi^2 \propto c^{-0.5} \tag{7.29}$$

The blobs are static and dynamical units:

$$s_{blob} = \frac{s}{g} = \frac{1}{6\pi\eta\xi}$$

At $c^*$ and below, in the dilute regime, we reach the single chain regime:

$$s = \frac{N}{6\pi\eta R_F} \sim N^{0.4}$$

Direct measurement of $s$ can be achieved by ultracentrifugation (Figure 7.50).

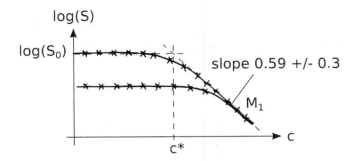

FIGURE 7.50 Sedimentation coefficient $s$ of polystyrene in good solvent (benzene) vs. concentration $c$ for different polymer molecular weight M measured by ultracentrifugation. Above a threshold concentration $c^*$, $s$ becomes independent of M.

### 7.7.2.2 Cooperative Diffusion: $D_{coop}$

Consider a longitudinal mode of wave vector $q$ of monomer concentration $c$ shown in Figure 7.51.

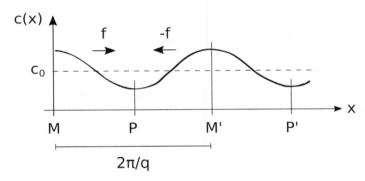

FIGURE 7.51 Breathing mode of a polymer solution.

$$c(x,t) = c + \delta c_q(t)\cos(qx).$$

This modulation of concentration gives rise to a force per monomer $f_m = -\dfrac{\partial\mu}{\partial x}$ and a flow $J_p$ of monomers:

$$J_p = cv = cs\left(-\frac{\partial\mu}{\partial x}\right) = -cs\frac{\partial\mu}{\partial\pi}\frac{\partial\pi}{\partial c}\frac{\partial c}{\partial x} \tag{7.30}$$

Using the hydrodynamic relationship $\dfrac{\partial\mu}{\partial\pi} = \dfrac{1}{c}$ leads to

$$J_p = -s\frac{\partial\pi}{\partial c}\frac{\partial c}{\partial x} = -D_{coop}\frac{\partial c}{\partial x} \tag{7.31}$$

Inserting in Equation 7.31 the sedimentation coefficient $s = \int (1/6\pi\eta r) g(r) dr$ and the osmotic compressibility $\chi = (\partial c/\partial \Pi)$ related to the pair correlation function $g(r)$, $\chi = (1/k_B T) \int g(r) dr$, leads to

$$D_{coop} = \frac{k_B T \int \frac{1}{6\pi\eta r} g(r) dr}{\int g(r) dr} = \frac{k_B T}{\eta \xi} \sim c^{0.75} \tag{7.32}$$

Where we have used the Ornstein–Zernike approximation $g(r) \cong (1/a^2 r) e^{-r/\xi}$.

Adding the conservation equation $(\partial c/\partial t) + (\partial J/\partial x) = 0$ leads to a relaxation rate given by:

$$\omega(q) = D_{coop} q^2 \tag{7.33}$$

This holds only at small $q$. The dynamical scaling hypothesis leads to:

$$\omega(q) = D_{coop} q^2 \Omega(q\xi)$$

With $\Omega(x) \to 1$ for $x \to 0$ and $\Omega(x) \to x^Z$ for $x \to \infty$. The exponent $z$ is derived from the condition that at large $q$, the dynamics are independent of $R_F$. This leads to a dynamical exponent $z = 3$.

At large wave vectors $q\xi > 1$, $\omega(q) \sim q^3$.

Transient *gel* regime: $\omega T_r < 1$ defines the hydrodynamic regime, where the long chains are relaxed, and $D_{coop}$ is called the "isotherm" diffusion coefficient. On the other hand, if $\omega T_r > 1$, the chains are not able to relax the entanglements, and the solution does not behave like a liquid but like a gel. The measure of $D_{coop}$ "adiabatic" in this regime gives the same scaling behavior.

The cooperative modes have been determined by inelastic light scattering experiments by Adam and Delsanti [6]. The value 0.68 obtained experimentally for the critical exponent of $D_{coop}$ vs. $c$ is smaller than the theoretical value 0.75. The cross-over between the dilute and the semi-dilute regimes occurs at a concentration $c^* \sim M^{-0.81}$. This molecular weight dependence is quite close to the $M^{-0.8}$ power law dependence predicted by the theory.

### 7.7.2.3 Dynamics of Gels

Figure 7.51 represents a polymer gel immersed in a good solvent. The deformation of the gel is characterised by the displacement $u(x)$ of the crosslinks.

A swollen gel can be represented as a mesh whose size is given by the Flory radius (Figure 7.52). The dynamical equation for the fluctuation of a gel is given by the balance of elastic and friction forces. For a deformation $u(x)$, this equation can be written as:

$$E \frac{\partial^2 u}{\partial x^2} = 6\pi\eta R_F \frac{c}{N} \frac{\partial u}{\partial t},$$

where $E = (c/N) k_B T$ is the elastic modulus, $6\pi\eta R_F$ is the friction per blob, and $c/N$ is the density of blobs.

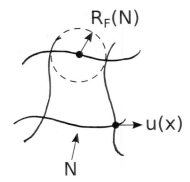

**FIGURE 7.52** Schematic representation of model swollen gels immersed in a good solvent. At equilibrium, the monomer concentration $c$ is equal to $c^* = N/R_F^3$, with $N$ the number of monomers between crosslinkers and $R_F = N^{3/5}a$ the Flory radius. The deformation of the gel is characterized by $u(x)$.

It leads to the same expression for $D_{coop}$:

$$D_{coop} = \frac{k_B T}{6\pi\eta R_F}$$

7.7.2.3.1 Experiments    A model polymer network has been prepared by anionic block copolymerization of styrene with divinylbenzene (DVB) in the proportion of four DVB per living polystyrene. When they are immersed in a good solvent (benzene), they swell. At equilibrium, each PS chain has a size $R_F = N^{3/5}a$. This is known as the $c^*$ theorem demonstrated by P.-G. de Gennes.

The cooperative diffusion constant, determined from light scattering experiments on polymeric networks swollen in a good solvent, varies with equilibrium concentration according to a scaling law similar to that observed in semi-dilute solutions [7].

### 7.7.2.4 Self-Diffusion $D_{self}$ in Semi-Dilute Solutions

The average distance between entanglements is the mesh size $\xi$ of the polymer solution. The test chain can then be considered as made of a succession of blobs of size $\xi$ as shown Figure 7.48, and its length is $L_t = (N/g)\xi$. If $g$ is the number of monomers per blob, $g = c\xi^3$, the mobility of the chain along the tube is $s_t = s_{blob}/(N/g)$, where $s_{blob} = k_B T/\eta\xi$ is the mobility of a blob. The diffusion coefficient along the tube $D_t$ is related to $s_t$ by an Einstein relation $D_t = k_B T s_t$. In a way similar to the melt case, we define the reptation time $T_r$ as the time to diffuse over a length $L_t$: $D_t T_r = L_t^2$. But the center of mass moves only on a distance equal to $R(c)$, which defines $D_{self}$:

$$D_{self} T_r = R^2(c) \qquad D_{self} = \frac{k_B T}{\eta\xi}\left(\frac{g}{N}\right)^2$$

where $R(c)$ is the radius of the ideal chain of blobs $R(c) = (N/g)^{1/2} \xi$. $T_r$ is the time it takes for the chain to completely renew its configuration, i.e. move the center of mass over a distance $R(c)$. The self-diffusion of entangled chains in good solvent is then:

$$D_{self} = D_0 N^{-2} \Phi^{-7/4},$$

which is in good agreement with experimental results of Leger and Rondelez by FRS (Section 7.6 and reference [6] therein).

$$D_{self} \sim c^{-1.7 \pm 0.1} N^{-2 \pm 0.1}$$

### 7.7.2.5 Dynamical Exponent z

We give in Table 7.3 values of $z$ obtained by a large variety of experimental techniques.

TABLE 7.3   Experimental Determination of the $z$ Exponent

| $\tau_z \sim R^z$ | $z_{theo} = 3$ |
|---|---|
| Diffusion $D$ | 2.91 |
| Mobility (sedimentation) | 2.88 |
| $\Delta \omega_q$ | 2.85 |
| Viscosity | 2.82 |
| Mechanical properties | 2.78 |

### References

1. P.E. Rouse, *J. Chem. Phys.*, 21, 1272 (1953).
2. B.H. Zimm, *J. Chem. Phys.*, 24, 269 (1956).
3. P.-G. de Gennes, *Scaling Concepts in Polymer Physics*. Cornell University Press, 1979.
4. M. Adam, M. Delsanti, *Macromolecules*, 10, 1229 (1977).
5. C. Destor, F. Rondelez, *J. Polym. Sci.*, 17, 527 (1979).
6. M. Adam, M. Delsanti, *J. Phys.*, 37, 1045 (1976).
7. J.P. Munch, S. Candau, *J. Phys.*, 38, 971 (1977).

# CHAPTER 8

# Soft Matter in Everyday Life

## 8.1 SELF-CLEANING SURFACES: FROM LOTUS LEAF TO SHARK SKIN

Let us compare the image of the hull of a boat after a few months at sea with that of the skin of a shark or a lotus leaf (Figure 8.1). How do sharks and plant leaves not get dirty?

FIGURE 8.1    (A) The "dirty" surface of a boat's hull. (B) Self-cleaning surfaces of a shark (copyright Shutterstock). (C) A lotus in Central Park, New York. (Photo FBW.)

Shells readily stick to boat hulls, and coverage of the hull with algae and mussels can become almost complete after a few months. Chemical treatments are ineffective because mussels and other shellfish adhere very strongly to all types of surfaces, both hydrophilic and hydrophobic, and are able to resist storms. They secrete glues that also inspire the glue industry. With marine transport, which has taken on enormous importance, these problems of marine fouling are the subject of a great deal of research (see for instance the works by Jan Genzer [1] at the University of North Carolina, in the United States).

More generally, we will describe here how plants and animals manage to generate antifouling surfaces. Nature provides many examples of mechanisms to control fouling. Super hydrophobicity (i.e. extreme non-wetting) and its implications are presented. These natural types of defense emerging from evolution can be copied (biomimetic surfaces) or adapted (bio-inspired surfaces) to solve fouling problems on artificial structures, such as glass skyscrapers or the Louvre pyramid.

This research field has been very active in the past two decades and focuses on new physical mechanisms to prevent and control fouling. With the advent of new nanotechnologies for 3D replication and printing, which allow natural surfaces to be replicated and new structures to be invented, the characterization and modeling of nano- and micro-scale systems is now within reach. In addition, mechanical deformations of surfaces generated by stretching and inducing wrinkles (e.g. human skin wrinkles) have made it possible to assemble new structures and understand important physical phenomena.

### 8.1.1 Natural Antifouling Mechanisms

Most plants originating from marine life are equipped with a membrane that protects them against the hostile external environment and first and foremost allows them to fight against desiccation out of the water. During their development plant skins became more diversified. Several functional properties have emerged, such as water movement control, surface wetting properties, resistance to pollutants and pathogens, pheromone secretion, radiation resistance, photosynthesis control, and improvement of their mechanical and thermal properties.

The epidermal cells of plants have an outwardly stratified structure (Figure 8.2): beyond the plasma cellulose membrane, the cuticle, a layer composed of cutin and cellulose, serves as a barrier to perspiration and as a structural element of the cell. Then, in direct contact with the environment, more or less thin films of wax cover the cuticle and may present secondary structures like crystals of various forms: hair, plaques, papillae, etc. The variable structures of the cuticle have several purposes: control of water supply, wetting properties, etc. These structures are studied from the microscopic scale of groups of cells to the nanoscopic scale of the surface of a cell. Indeed, the wettability properties of plant surfaces are controlled by the structures adopted at all these scales. They also depend on a simple physicochemical factor: the properties of the wax covering the cuticle (2D films, 3D crystals).

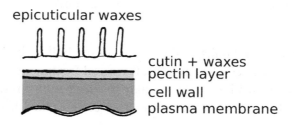

FIGURE 8.2    Drawing of the structure of the outer surface of a plant leaf.

### 8.1.2 Wetting on Rough Surfaces

Four wetting regimes can result from these properties: hydrophobic, super-hydrophobic, hydrophilic, and super-hydrophilic, which are optimal to allow the survival of each plant in its environment. These wetting behaviors are characterized by the measure of contact angle and hysteresis (difference between advancing and receding angles – Section 4.4), assessed by the critical angle of inclination of the substrate that produces the movement of a deposited droplet.

The waxes covering plants are a hydrophobic material, but the structure of the surface makes it possible to vary this behavior: the contact angle defined by Young's relation considers ideal surfaces without roughness. The Wenzel equation, which assumes full contact between the drop and the surface, provides a correction to take into account the roughness $r$ of the material, which amplifies the hydrophilic or hydrophobic behavior (Figure 8.3). According to the Wenzel law (1936): $\cos\theta_W = r \cdot \cos\theta_E$, where $\theta_W$ is the apparent contact angle and $\theta_E$ is the Young contact angle. If $\theta_E < \pi/2$, $\theta_W < \theta_E$: the rough substrate is more hydrophilic; if $\theta_E > \pi/2$, $\theta_W > \theta_E$: the substrate becomes more hydrophobic and even superhydrophobic for $\theta_W = \pi$.

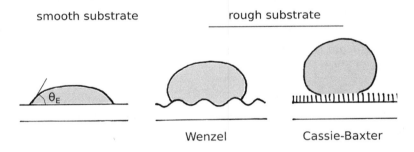

FIGURE 8.3   Different wetting regimes, depending on the roughness of the surface.

### 8.1.3 Self-Cleaning Processes

Hydrophilic behavior is necessary for the petal to allow access to pollinating insects. In petals, cells are only coated with a 2D film of wax or secretory hairs and glands.

Super-hydrophilic behavior is observed in plants with high water requirements, such as sphagnum moss or roots: their walls are porous or equipped with papillae or hair to favor water absorption. The pore size is adapted to the available source of water: dew, mist, or rain.

A simple hydrophobic behavior prevents the formation of a water film on leaves but would also hinder the transport of $CO_2$ required for photosynthesis. On leaves, convex cells or cells with cuticle structures covered by wax or 3D crystals are usually observed.

The super hydrophobic behavior observed for the leaves of some plants (lotus, taro, nasturtium) is due to stud-like structures present on the surface (convex cells or cells with papillae) topped by 3D crystals of epicuticular waxes (Figure 8.2). The surface presented to the drop only gives rise to very limited contact with water, with air being trapped under the surface irregularities. The observed behavior follows Cassie's law (Figure 8.3). The contact angle is greater when the drop is in contact with air.

This behavior provides the leaves of these plants with self-cleaning properties against particles larger than the spacing between studs, which are carried away by water flowing on the surface of the leaf. In addition, some aquatic ferns have a hairy surface that has super hydrophobic properties and allows them to float by maintaining a gaseous (air) film. This phenomenon is also seen in the animal kingdom: the aquatic spider makes a large air bubble that serves as a house. Ducks are super hydrophobic and never get wet due to the presence of hydrophobic nanoparticles on their feathers. There have been attempts to make self-drying shampoos in order to allow one to get out of the pool with dry hair, but it proved impossible to use nanoparticles because of other undesired effects on the hair.

### 8.1.4 Engineered Self-Cleaning Surfaces

Understanding these behaviors has made it possible, particularly in the case of the lotus effect, to engineer super hydrophobic surfaces: self-cleaning, or "not wet" even under water, which allows the reduction of hydrodynamic drag during fluid movements or transport. Such shark skin suits that increased the performances of swimmers were banned at the Olympic Games.

Thanks to evolution, nature offers a myriad of solutions to the problems faced by plants and animals everywhere on Earth. By drawing inspiration from nature, materials science, and in particular surface physicochemistry, can lead to innovative solutions to various problems, ranging from the maintenance of public buildings to maritime transport and even to microfluidics.

### References

1. J. Genzer, K. Efimenko, *J. Bioadhesion Biofilm Res.*, 22, 339–360 (2006).
2. K. Koch B. Bhushan, W. Bathlott, *Soft Matter*, 4, 1943–1963 (2008).

## 8.2 HYDROGEL PEARLS IN MOLECULAR CUISINE AND CELL BIOLOGY

### 8.2.1 Spherification and Molecular Cuisine

Many raw foods (like milk) or others, which are processed by simple cooking operations (like mayonnaise), are emulsions (Section 4.1). The droplets of oil stabilized with egg yolk lecithin are sub-millimetric in size and very numerous, which gives the mayonnaise its almost solid appearance. But divided systems can also be found in gourmet cuisine. In the early 2000s, a fashion in molecular gastronomy was the use of hydrogels through the so-called spherification technique. This technique was patented by William Peschardt, who worked as a chemist at Unilever, in 1942 for culinary applications [1] but only developed and popularized in 2003 by Ferran Adrià, the chef of the Spanish restaurant *El Bulli*. Unlike gelation that can be achieved with agar-agar for example, spherification consists of a partial gelation that encapsulates drops of liquid food in a solid membrane. To do this, chefs use two additives: sodium alginate and calcium lactate.

### 8.2.2 Formation of an Alginate Hydrogel

Sodium alginate is a biocompatible polymer extracted from brown algae. It is more precisely a polysaccharide composed of two types of monomers, mannuronic acid (M) and guluronic acid (G). The presence of carboxyl groups (COO⁻) on residues G and M renders the alginate highly negatively charged. As a consequence, divalent cations, and in particular calcium, can induce crosslinking bridges by complexing some $COO^-$ within the same alginate chain or between two molecules. For conformational reasons, the interaction affinity is greater between calcium and monomer G than between calcium and monomer M. This leads to gels that have an "egg box" structure, such as sketched in Figure 8.4A.

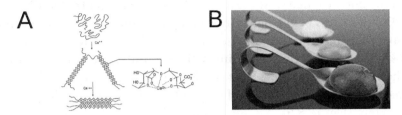

FIGURE 8.4   (A) Sketch of alginate crosslinking by calcium ions and magnification on the chemical structure showing calcium bridges within the physical gel. (B) Flavor spheres reconstituted from alginate and strawberry or mango juice. (Copyright Shutterstock.)

To form pearls of flavor, we simply mix a dilute solution of alginate with the food to be encapsulated in the form of coulis or puree. By loading the solution into a syringe and dropping small drops in a calcium bath, crosslinking of the superficial layers is fast enough that the drop does not fall apart during the impact. Calcium ions then enter from the solution into the droplet of alginate and food. By neglecting any convection and assuming that the transport of calcium ions, and thus the crosslinking kinetics of the drop, are governed by diffusion, the time required to reach complete gelation for a drop of 2 mm in diameter (which corresponds to the size of water-based drops that fall under their own weight – Section 4.3) is $t \sim R^2/2D_{calcium}$, where $R$ is the radius of the drop and $D$ the diffusion coefficient of calcium ions, $D = 10^{-9}$ m²/s, i.e. $t \sim 1$ h. The trick is to remove the beads from the calcium bath after a few minutes to ensure that the core of the pearl remains liquid. It is indeed much more pleasant from a gustatory point of view to make a thin shell of alginate burst between his (or her) teeth, allowing the release of the encapsulated flavors rather than having the feeling of a flavored chewing gum. Larger drops can also be made using a spoon, and strawberry or mango flavored balls can be reconstituted (Figure 8.4B). To keep a liquid core, the optimal incubation time is two minutes and 30 seconds.

Before embarking on this culinary adventure like Ferran Adrià, it is necessary to insist on several points that make this technique delicate. First, the liquid core, which contains alginate molecules, eventually undergoes gelation over time due to remaining traces of calcium. These flavor beads cannot be conserved for typically more than 48 hours. Second, the control of the thickness of the alginate membrane remains fairly empirical and therefore requires a lot of trial-and-error cycles. Finally, these pearls are handmade, which limits their production. We can mention that the restaurant of Ferran Adrià has closed following lawsuits organized by some of his colleagues who considered that these additives though natural had, when ingested in large quantities, harmful effects on health (vomiting and diarrhea). To limit the amount of alginate, other Grand Chefs like Thierry Marx have sought to change the technique of formation of these beads. In collaboration with a chemist from the Ecole Supérieure de Physique et Chimie Industrielles in Paris, Jérôme Bibette, Thierry Marx has developed flavor pearls whose liquid heart does not contain alginate at all by using a micro-/milli-fluidic device developed in research laboratories. The production principle is based on the coextrusion of two liquids in a system of coaxial capillaries [2]. The food suspension is injected into the central capillary and the alginate solution injected into the outer capillary. At the exit of the double concentric needle, the drops are composite. The thickness of the outer layer of alginate solution can be controlled by the ratio of capillary sections and the flow rates. The lower the flow rate of alginate with respect to the food suspension flow rate, the thinner the outer alginate layer. The drops are then detached when they have reached a limit size fixed by the capillary length, that is the ratio of the surface tension to the gravity, and they fall dropwise into a gelling bath of calcium chloride or lactate (Figure 8.5). As crosslinking of the alginate is almost instantaneous, thin shells of alginate trapping the food suspension are formed.

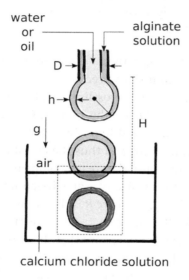

FIGURE 8.5   Fabrication principle of flavor pearls with liquid core. (Adapted from [2].)

The innovation brought by Marx and Bibette was to manage the impact of the drop on the surface of the calcium bath. To avoid the impact breaking the shell before complete gelation, the solution consists in (i) adding a trace of surfactant onto the surface of the calcium bath, which reduces the surface tension of the solution and thus the energy required for the passage of the interface, and (ii) adding surfactants into the alginate solution. This effect is more subtle: the precipitation of the surfactant in contact with calcium ions gives the drop a transient increased stiffness, which plays a protective role against the turbulence caused by the impact [2].

### 8.2.3 Application to Tissue Engineering and Oncology

Alginate beads have applications beyond the culinary field. Recently, the food suspension has been replaced by cancer cells [3]. As the hydrogel is permeable, with a pore size of the order of 20 nm, all the nutrients (proteins and oxygen) required for cell division can diffuse freely through the capsule and allow proliferation of the encapsulated cells. Since alginate is by nature cell-repellent, cells form cellular aggregates, called multicellular spheroids (Figure 8.6A), which are considered good *in vitro* models of micro-tumors. This opens up the possibility of high throughput screening of new drugs for chemotherapy. More recently, at the Optics Institute of Aquitaine and in collaboration with Aurélien Roux's laboratory in Geneva, Kévin Alessandri, the student who developed this technology, and his colleague Maxime Feyeux, a specialist in pluripotent stem cells, succeeded in encapsulating stem cells and transforming (or differentiating) them into neurons *in capsulo* [4] (Figure 8.6B). The applications are innumerable in the field of tissue engineering (organ-on-chip) and regenerative medicine (e.g. for the transplantation of dopaminergic neurons as a cell therapy for Parkinson's disease).

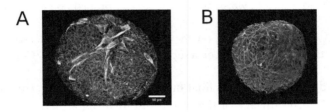

FIGURE 8.6 **(See color insert.)** Images of multicellular spheroids obtained by optical microscopy. Composition: (A) tumor (red) and stromal (green) cells. (Courtesy of Fabien Betillot.) (B) neurons (nuclei in blue, axons in magenta). (Courtesy of Kévin Alessandri.) The diameter of the spheroids is about 300 μm.

### References

1. W. Peschardt, Brevet US 2403547A (1942).
2. N. Bremond, E. Santanach-Carreras, L.-Y. Chu, J. Bibette, *Soft Matter*, 6, 2484–2488 (2010).
3. K. Alessandri, B.R. Sarangi, V.V. Gurchenkov, B. Sinha, T.R. Kiessling, L. Fetler, F. Rico, S. Scheuring, C. Lamaze, A. Simon, S. Geraldo, D. Vignjevic, H. Domejean, L. Rolland, A. Funfak, J. Bibette, N. Bremond, P. Nassoy, *Proc. Natl Acad. Sci. USA*, 110, 14843–14848 (2013).
4. K. AlessandriM. Feyeux, B. Gurchenkov, C. Delgado, A. Trushko, K-H. Krause, D. Vignjevic, P. Nassoy, A. Roux, *Lab Chip*, 16, 1593–1604 (2016).

## 8.3 WALK ON WATER LIKE SPIDERS AND LIZARDS

In his elementary theory of capillarity [1], Émile Duclaux writes in 1872:

> "Whenever a liquid has a free surface, it is there like enveloped in a contractile and very thin layer, constantly tense, resealing itself when it is broken, and of which one can get a clear idea, when everything is at rest, by comparing it with a very thin membrane of rubber enclosing the liquid."

Is this membrane strong enough for a man to walk on water? Do we necessarily need water skis and to be in motion? Is there an upper limit in weight to rest or even walk on the surface of the water?

### 8.3.1 Resting on Water: When Surface Tension Counteracts Weight

There is a classical experiment for children that can be found on the internet which consists of depositing a paper clip on the surface of the water (possibly with the help of a small piece of paper to facilitate the landing) and to make it sink by adding a trace of dishwashing liquid that lowers the surface tension of the water (Figure 8.7a).

paperclip      water strider      droplets on a lotus leaf

FIGURE 8.7  (A) A paper clip floating on the water. (B) A water spider walking on the water. (C) A drop of water that does not spread on a super-hydrophobic surface. (Copyright Shutterstock.)

Similarly, gerrids or water spiders float on water with their long, thin legs (Figure 8.7b). Although deforming the surface of the water and forming a meniscus, the weight of the spider is compensated by the surface tension forces $\gamma p$ where p is the perimeter of the legs; hence it is important for the spider to "make the big gap," that is to say to lay his legs horizontally. By taking a diameter of 100 μm and a length of about 2 cm for the legs, we find that the force due to the surface tension is of the order of 5 mN (with $\gamma = 72$ mN/m) while its weight $P = mg$ is only of the order of 1 mN (for a mass m of 100 mg). This effect, which makes him float despite a higher density than water, is accentuated by the existence of many small hairs along the legs. Indeed, not only does the waxy material that covers these hairs contribute to increase the hydrophobicity, but roughness also plays a decisive role in making them super-hydrophobic [2] (Section 8.1). This is a situation of wetting described by Cassie, corresponding to the behavior of fakir drops. The minimization of surface energy involves the creation of air pockets rather than having intimate liquid–solid contact throughout the hairs. The same phenomenon applies to lotus leaves and surfaces covered with carbon black (Figure 8.7c).

### 8.3.2 Movement of Light Water Spiders

How do water spiders move? They beat their pair of intermediate legs in the manner of the motion of oars in a boat. According to Newton's third law, there is a transfer of momentum to water (and air, but whose contribution is neglected because its density is much lower than that of water). From filming the insects on the surface of water, it appeared that waves were moving in the opposite direction to the movement of the insect. These capillary waves would naturally be the means of providing this momentum. The (phase) velocity of these waves depends on the acceleration of the gravity g, the surface tension $\gamma$, and the density of the water $\rho$. A detailed calculation [3] shows that these capillary waves only exist for a phase velocity higher than $c^{\star} = \left(4g\gamma/\rho\right)^{1/4}$ ~23 cm/s for the water–air interface. Thus, to produce capillary waves that make it move forward, the water spider must move its legs at a speed greater than 23 cm/s or turn them at about 50 turns per second! This is apparently impossible to achieve by the baby water spider. And yet, these also move. This paradox has been solved by John Bush and his PhD student, David Hu from MIT [4], who are specialists in animal locomotion [5]. By recording spider babies with fast cameras and tracking scattered particles dispersed in water to visualize fluid flows, the American group showed that the momentum transfer was mostly associated with the emission of small swirls. The process is similar to that which allows pigeons to fly or fish to swim. In the case of the spider, the vortex is created at the water–air interface [6] (Figure 8.8).

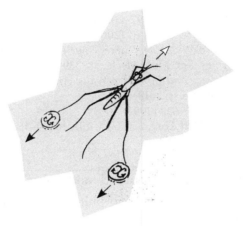

FIGURE 8.8    Illustration of the swirls generated during the movement of a water spider on the surface of water.

### 8.3.3 Movement of Heavy Lizards

For all heavier animals, characterized by a Baudoin number $P/\gamma p > 1$ (where P is the weight and $\gamma p$ the surface tension force), the above-mentioned arguments are no longer valid. Indeed, the weight of a man of 100 kg would be counterbalanced by a capillary force only if the perimeter of his foot was about 5 km! In the less extreme case of a small lizard of 100 g, its feet should be about 50 cm. Yet basilisk lizards (also known as Jesus lizards) walk on water. The explanation for their flotation lies in their movement. The walk, or rather the run, of a basilisk lizard on water is indeed composed of two main phases: a vertical thrust (generated by the slap of the foot on the water surface) and a horizontal thrust (generated by a horizontal attack) [7]. During the foot-strike phase, the associated momentum transfer corresponds to the product of the virtual mass associated with the foot by the speed with which this virtual mass is accelerated, that is to say about the speed of the foot at the impact. The virtual mass is derived from the volume of the air cavity created by the impact. The speed is measured using a fast camera. The force is obtained by dividing the momentum transfer by the duration of the contact and is about 0.3 N. During the phase of underwater attack, the hydrodynamic drag is proportional to the square of the speed and can be estimated to about 1 N, which is sufficient to support the weight of a 130 g lizard. The dependence of the drag with the square of speed makes us understand why it is advantageous to run fast!

192 ■ Essentials of Soft Matter Science

By adapting this approach for human beings, whose feet are small in proportion to their height compared to the lizard, one would find that the world record holder in the 100 m track, who runs at around 10 m/s, should be running at least three times faster to have a chance not to sink. A trick to achieve this, which has nothing to do with the capillarity, would be for example to replace water with a suspension of corn starch whose shear-thickening properties make it behave like a viscous liquid (honey) when mixed or stressed slowly and as a solid when the stress is sudden. It is then possible to run, even slowly, on a bath of corn starch provided you do not stop!

References

1. M.E. Duclaux, *J. Phys. Theor. Appl.*, 1, 197–207 (1872).
2. M. Callies, D. Quéré, *Soft Matter*, 1, 55–61 (2005).
3. P.-G. de Gennes, F. Brochard-Wyart, D. Quéré, *Gouttes, bulles, perles et ondes*. Belin, 2005.
4. J.W.M. Bush, D.L. Hu, *Annu. Rev. Fluid Mech.*, 38, 339–369 (2006).
5. D. Hu, *How to Walk on Water and Climb Up Walls: Animal Movement and the Robots of the Future*. Princeton University Press, 2018.
6. M. Dickinson, *Nature*, 424, 621–622 (2003).
7. T. Hsieh, G.V. Lauder, *Proc. Natl Acad. Sci. USA*, 101, 16784–16788 (2004).

## 8.4 WALK UP WALLS LIKE GECKOS

### 8.4.1 Performances of Geckos and Discarded Mechanisms

Who has never dreamed of climbing the walls of buildings with bare hands like Spiderman? We may be on the way to fulfilling this dream, and this has been made possible after understanding the locomotion of an animal. Aristotle, in the 4th century BC, was fascinated by the woodpecker that could climb trees, even upside down, in the same way as the gecko lizard (Aristotle, *The History of Animals*). The woodpecker and the squirrel get their claws into the tree trunk to increase friction and adhesion forces and compensate for gravity. On a smaller scale, the same type of feat is performed by cockroaches. If it is less surprising, the mechanism is the same because the legs of these insects are also provided with claws. Yet all these clawed animals are powerless when it comes to climbing polished walls of buildings, unlike the gecko. Is the gecko just a snail or a four-legged slug secreting a thixotropic sticky mucus, solid at rest and liquid when sheared? Or a big ant that secretes liquids and uses capillary adhesion (Sections 4.2 and 4.3) to adhere to smooth surfaces (and claws for rough surfaces)? No, because on the one hand the gecko leaves no trace of liquid on its way, and on the other hand, it can, contrary to the snail, adhere as well on hydrophilic surfaces (like clean glass) as on surfaces hydrophobic (like a non-stick frying pan).

### 8.4.2 Hierarchical Adhesion Structures of Geckos

The mystery of adhesion and locomotion of the gecko has been elucidated by the group of Kellar Autumn at the University of Portland, United States. The American researchers started from a detailed observation of the legs of the gecko reported in 1965 by R. Ruibal and V. Ernst [1]. The gecko, with an average mass of 50 g, has four legs (with an area of 1,00 mm² each) consisting of five fingers. The finger pad is a highly branched structure. The fingers are lined with strips of setae. These setulae, 1,00 μm long and 5 μm in diameter, have a density of 10,000/mm², or one million per leg. These setulae branch themselves into 1,00–1,000 fibrils terminated by a spatula of 200 nm side. This hierarchical structure (Figure 8.9) allows the gecko to withstand a tensile force of about 20 N, i.e. 40 times its weight.

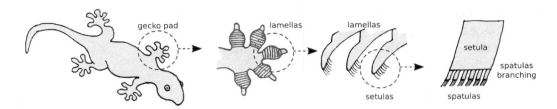

FIGURE 8.9   Gecko adhesive system showing a hierarchical structure.

The existence of spatulas suggested that they could act as suction cups in a suction phenomenon, thus allowing considerable adhesion forces. However, in 1934, a German researcher, Wolf-Dietrich Dellit [2], discarded this hypothesis by showing that the adhesion of a freshly cut gecko leg was not altered under vacuum. Without knowing the exact structure of the gecko legs, Dellit suggested an electrification mechanism by contact between the legs and the surface, giving rise to electrostatic interactions that can also be created when rubbing a cat's skin or a piece of tissue on a rod of glass. In 1949, Dellit ruled out this new hypothesis after showing that the adhesion was not modified under ionized air irradiation that serves to suppress electrostatic interactions [3].

In 2002, the Kellar Autumn group tested the hypothesis that sub-micron spatulas may interact with surfaces via Van der Waals forces [4]. In this case there is no need to create, per se, electrical charges, since Van der Waals interactions are dipolar by nature. But, as mentioned in Section 2.3, these interactions, which are always present in nature, are very short-range ($E \sim 1/d^6$ at the molecular scale and $\sim 1/d^2$ per unit area between two surfaces). They are therefore effective only if the separation distance is nanometric. Indeed, when we touch our palms with each other, they do not remain stuck, because the roughness of the skin is such that the actual "atomic" contact zone only represents a tiny fraction of the surface of the hand, which is insufficient to mediate strong adhesion. The advantage of the gecko lies in the fact that the flexibility of the setulae allows the spatulas to optimize the area of actual contact with the surface, which leads to an amplification of the adhesion energy.

### 8.4.3 Measurement of Attachment Force

Direct measurements of the detaching force of the gecko legs show that the two front legs can withstand 20 N. The corresponding area of 200 mm² therefore comprises about two million setulae, and $\sim 10^9$ spatulas. Assuming that the spatula is a half-sphere of radius $R = 100$ nm, the force $f_{spatula}$ to be applied to detach a spatula is, according to a conventional theory of adhesion (the Johnson–Kendall–Roberts theory) of the order of $W \times R$, where $W$ is the separation energy density. If the interaction is of the Van der Waals type, it is of the order of 50 mJ/m², and $f_{spatula} \sim 5$ nN. Then the total force for both feet is $f_{total} = N_{spatula} \times f_{spatula}$ ~ 5 N, which is slightly lower than the measured force but of the same order of magnitude, which shows that the assumption that the adhesion of the gecko leg is dominated by Van der Waals interactions is consistent.

These considerable attachment forces finally raise the question of how geckos manage to move so quickly, since they take less than 15 ms to detach their feet. In fact, the setulae are curved. Due to this asymmetry the breaking force becomes a function of the relative angle between the setulae and the surface, which allows the lizard to run.

### 8.4.4 Biomimetic Gecko

These experiments have initiated many biomimetic approaches with the prospect for us to finally become Spiderman. For instance, the StickyBot was created on the principle of the gecko at Stanford University [5], and a super-powerful Geckskin tape has been marketed from the work of a team from the University of Amherst [6].

### References

1. R. Ruibal, V. Ernst, *J. Morphol.*, 117, 271–293 (1965).
2. W.-D. Dellit, Jena Z. *Nature*, 68, 613–656 (1934).
3. W.-D. Dellit, *Deut. Aquar. Z.*, 2, 56–58 (1949).
4. K. Autumn M. Sitti, Y.A. Liang, A.M. Peattie, W.R. Hansen, S. Sponberg, T.W. Kenny, R. Fearing, J.N. Israelachvili, R.J. Full, *Proc. Natl Acad. Sci. USA*, 99, 12252–12256 (2002).
5. S. Kim, M. Spenko, S. Trujillo, B. Heyneman, D. Santos, and M. R. Cutkosky, Robotics, *IEEE Transactions on*, 24(1), 65–74 (2008).
6. M. D. Bartlett, A. B. Croll, D. R. King, B. M. Paret, D. J. Irschick, and A. J. Crosby. Advanced Materials, 24, 1078–1083 (2012).

## 8.5 CAPILLARITY AT THE SERVICE OF PLANTS

### 8.5.1 Sap Rise and Cavitation in Trees

Water and nutrients routing to the leaves of a plant and the return of photosynthesized sugars to all parts of the plant are reminiscent of animal blood circulation through the network of veins and arteries. However, unlike the animal vascular system, the processes involved in plants are purely passive. One of the wonders of trees is that they reach heights of nearly 100 m (Figure 8.10). Immediately, using Jurin's law (Section 4.3), one finds that the radii of xylem capillaries should be as small as $r \approx (2\gamma/\rho g) = 200$ nm for capillarity to be effective. However, since the xylem conducting vessels have a minimum radius of 20 μm, other mechanisms should be at play. Osmosis allows penetration into the roots and rise up to $h \approx (cRT/\rho g) = 20$ m maximum (for high sugar concentrations of the order of 30 g/l and a molecular weight of 300 g/mole). In fact, the aspiration of sap is primarily produced by the evaporation at the level of the leaves. However, the expression of the hydrostatic pressure $P(H) = P_0 - \rho g h$ shows that, for atmospheric pressure at ground level, the vacuum is reached at $H = 10$ m above ground level. Beyond this height, water is therefore in a metastable state of negative pressure. The sap must rise while maintaining its cohesion to avoid cavitation phenomena (embolisms). Research remains active in this field.

FIGURE 8.10   Capillarity and sap rise in a tree.

## 8.5.2 Surface Tension Propulsion of Fungal Spores

If capillarity is not so helpful for trees, we will see that it plays a preponderant role in the reproduction of certain plants. The dispersion of spores or pollen in mushrooms and plants is indeed crucial, because it is necessary to move the material required to ensure the next generation. As for fungi, spores are rather elongated particles of about ten microns in length. They grow on the lamellae of mushrooms. The challenge is to eject them out of the lamellae, with a horizontal trajectory first to make them arrive at the halfway point then free fall (Figure 8.11A) while subjected to convection of air (or wind), which can spread them away. Several million spores can be ejected in one hour. Spores then have the peculiarity of being very resistant to difficult weather conditions (frost, humidity, drought).

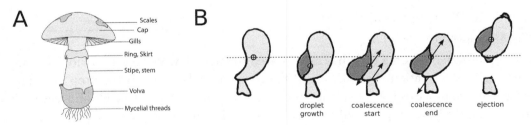

FIGURE 8.11   (A) Spores fallen from mushroom lamellae (copyright Shutterstock). (B) Zoom on a spore. Condensation of the drop at the base of the spore followed by the ejection of the spore. The arrows indicate the forces involved. The force exerted by the drop and the spore gives rise to a global force exerted at the center of mass by the drop–spore complex. The reaction of the "support" (hilar appendage) finally reaches the level of the breaking force. (Adapted from [2].)

This mechanism has been recently elucidated. The ejection occurs through the birth of a water droplet by condensation at the base of a spore. More precisely, this condensation is induced by the exudation of solute (mannitol) on the hilar appendix [3]. In a humid atmosphere, this drop grows and ends up touching the spore, which is wetting. Spreading of the drop over the spore occurs at a characteristic speed that results from the balance between capillarity and inertia ($\rho V^2 \approx \gamma / R$), i.e. $V \approx \sqrt{\gamma/\rho R}$. With $R \approx 5\ \mu m$ for the radius of the drop, $V$ can reach 10 m/s. The energy gain associated with the surface area reduction ($\approx \gamma R^2$) is transformed into kinetic energy release ($\approx \rho R^3 V^2$) to the drop, then to the spore. This idea, which had been first proposed by Turner and Webster [4], was validated, made precise, and quantified by X. Noblin, a researcher in Nice Sofia Antipolis, while he was doing a postdoctoral internship in the laboratory of Jacques Dumais at Harvard, United States [2].

Using a high-speed camera that allows one to acquire images at a rate of 25,000 frames per second, X. Noblin and his colleagues recorded the ejection of individual spores. The energy transfer (between the coalescence of the drop and the propulsion of the spore) takes place in less than 4 μs. The conversion of this surface energy into kinetic energy leads to initial velocities of nearly 1 m/s, and the force exerted by the drop onto the spore (and vice versa) is of the order of 1 μN. As seen in Figure 8.11B, the propulsion sequence of the spore is reminiscent of the jump of a human being, who bends his legs and then propels himself. Here, the reaction of the ground plays the role of the coalescence of the drop and allows him to jump. Remarkably, the take-off speed of the best animal jumpers is also a few m/s. In their case, since the mass is much larger, so is the momentum, but friction in the air slows them down more than these micrometer-sized spores.

References

1. A. Buller, *Researches on Fungi*, vols. 1–7. London, UK: Longmans, Green and Company (1909 & 1950).
2. X. Noblin, S. Yang, J. Dumais, *J. Exp. Biol.*, 21, 2835–2843 (2009).
3. J. Webster, R. Davey, N. Smirnoff, W. Fricke, P. Hinde, D. Tomos, J. Turner, *Mycol. Res.*, 99, 833–838 (1995).
4. J. Turner, J. Webster, *Chem. Eng. Sci.*, 46, 1145–1149 (1991).

## 8.6 DROPLETS IN THE KITCHEN

### 8.6.1 Evaporation of a Drop

A drop of water left on a table has disappeared the next day. Water has obviously evaporated. This evaporation is accelerated if the room or table is hot. If the drop of water is now replaced by a drop of coffee, the coffee particles will settle on the table and leave a dry trace. Intuitively, we might think that we would find denser discs of coffee particles in the center and more dilute discs at the edges, because the volume of liquid in the center (in the vicinity of the dome of the drop), thus the number of particles, is higher than at the drop edge. In reality, we observe crown-shaped patterns (Figure 8.12A). Almost all particles have accumulated at the periphery of the initial drop. This effect was named the "coffee stain effect" and explained by Tom Witten at the James Frank Institute in Chicago in 2000 [1]. It occurs if the contact line of the drop is "trapped," which is generally the case on real surfaces that are not topographically smooth and chemically homogeneous at the atomic scale. As shown in Figure 8.12B, during evaporation, the volume of the drop decreases, but it is necessary that the liquid removed near the edge is compensated by a flow of centrifugal liquid in order to keep the same footprint (or surface of contact). This flow of liquid carries the coffee particles with it. Note that on the other hand, if the drop can retract as it evaporates, the deposition of coffee bean particles will be more homogeneous.

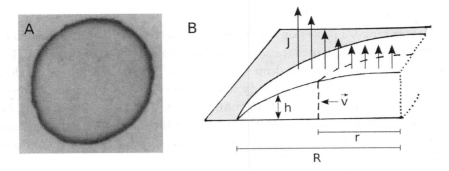

FIGURE 8.12  (A) Photograph of a coffee stain. (B) Drawing of an evaporating drop with trapped contact line. The steam flow $J(r)$ is higher in the periphery because the area is larger when moving away from the center $r = 0$. The height of the drop $h(r)$ decreases. But, in the periphery, the corresponding lost volume is less than that lost by evaporation. So, there is a flow of centrifugal liquid at the average speed to compensate for this volume loss.

## 8.6.2 Levitation of a Drop

Let us consider again a drop of pure water. What happens if a drop of water is placed on a very hot surface, at a much higher temperature than the boiling temperature of the water? One would expect instant evaporation or violent boiling accompanied by an explosion of the drop. However, that is not at all what happens. Everyone has already experienced this in their kitchen or dropping water droplets, voluntarily or not, on the glowing cooking plate. The drops keep a rounded shape and acquire great mobility: they start dancing on the plate. This phenomenon is called calefaction or the Leidenfrost effect, named after the German doctor who was interested in it in the 18th century [2]. A century later, Jules Verne also used this effect in his book *Michel Strogoff* (1876), in which his hero had his eyes exposed to a white-heated sword and simulated blindness before defeating his enemy. His tears vaporized as the blade approached and created an insulating and protective film of vapor. This is exactly what happens with the drop which is not in direct contact with the cooking plate but rests on a 100–200 μm thick steam cushion (Figure 8.13). This cushion, by evacuating itself, gives an erratic movement to the drop that dances on the plate. And the steam cushion is immediately renewed.

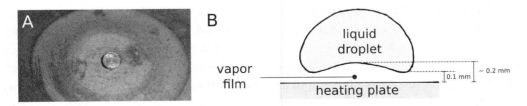

FIGURE 8.13  (A) Drop of water on a glowing cooking plate. (B) Drawing of drop levitation on top of a cushion of steam in the Leidenfrost effect.

### 8.6.3 Self-Propulsion of a Drop

In 2006, Heiner Linke, a professor at the Swedish University of Lund, deposited drops of alcohol (ethanol) on roof-shaped (or sawtooth) brass surfaces heated to 350°C. If the sawtooth pattern is asymmetrical, the drop is self-propelled in the direction of the downward slope of the teeth (Figure 8.14). Its steady state speed can reach 10 cm/s. The topography allows the steam flow to be "rectified" or funneled. Small particles dispersed in the steam are observed to follow the more gentle slopes before hitting the steeper slope and being evacuated to the side. This steam flow generates a viscous force that pulls the drop, hence the direction of movement.

FIGURE 8.14 Ethanol drop moving on a factory roof-shaped surface heated to 300°C by rectified Leidenfrost effect. (Extracted from [2].)

### 8.6.4 Running Drops

A totally different way of moving drops of liquid on a solid surface was proposed in 1995 by Thierry Ondarçuhu [3]. This experiment consists of taking drops of alkane in which silane molecule are dissolved. As discussed in Section 4.5, silanes are surface-active molecules, composed of a long hydrogenated or fluorinated chain and a hydrophilic head that reacts with glass silanol surface groups (Si–OH) to form covalent bonds. Thus, if a clean, hydrophilic glass surface is dipped into a silane solution, it becomes hydrophobic. When T. Ondarçuhu deposited a drop of alkane containing perfluorodecyltrichlorosilane molecules on a glass surface and then gave it a little push with a glass pipette, the drop began to move long distances in a self-supporting movement. Remarkably, by creating a "breathing pattern," i.e. by generating fog on the glass by the breath of our mouth, it condenses preferentially on the hydrophilic zones, which made it possible to reconstruct the trajectory of the drop and to verify that, by moving, the drop left silane molecules behind. It should be noted in the photograph of Figure 8.15A that the trajectory is a little irregular, because the drop is deflected by glass inhomogeneities, but it is strictly self-avoiding, i.e. the trajectory does not intersect. A precise calculation has been described by one of us (FBW [4]). We give here only a qualitative explanation of the underlying mechanism. Silane molecules graft onto the glass under the drop deposited and increase the contact angle. At the time of the flick, part of the drop "sees" a bare glass surface. Its contact angle is therefore smaller (Figure 8.15B). The capillary forces acting on the drop at the rear (point A) and front (point B) are not compensated. The drop moves in the direction of the highest solid–air interfacial energies. An order of magnitude calculation suggests that a drop containing 50 mMol/L of silane can move at least 10 meters before the effect is exhausted.

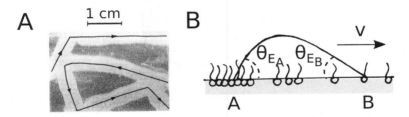

FIGURE 8.15 (A) Breath figure reproduces the trajectory of a "running drop" (T. Ondarçuhu). (B) Drawing shows the shape of the drop due to the difference in wettability.

## References

1. R.D. Deegan et al., *Phys. Rev. E*, 62, 756–765 (2000).
2. G. Lagubeau M. Le Merrer, C. Clanet, D. Quéré, *Nat. Phys.*, 7, 395–398 (2011).
3. F. Domingues Dos Santos, T. Ondarçuhu, *Phys. Rev. Lett.*, 75, 2972–2975 (1995).
4. F. Brochard-Wyart, *Langmiuir*, 5, 432–438 (1989).

## 8.7 FROM SOAP BUBBLES TO HURRICANES

### 8.7.1 The Rayleigh–Bénard Instability

Under gravity, water flows into a soap film that is held vertically. The film gets thinner in its upper part. It becomes a black film before bursting (e.g. by adsorption of a dust which serves as a nucleator). This drainage phenomenon can be slowed down in microgravity or compensated for if a water circulation is set up between the bottom and top of the film, but this is not simple to implement in practice. However, a fairly simple way to extend the life of a bubble, although completely counter-intuitive at first glance, is to heat it! The principle consists in using a hydrodynamic instability that is well known in geophysics, meteorology, astrophysics, and oceanography, namely the Rayleigh–Bénard instability. Rayleigh–Bénard cells shown in Figure 8.16 are observed when heating water in a pan. Since there is a temperature gradient between the cooking plate and the liquid surface, and because hot water is less dense than cold water, hot water molecules rise in the container while molecules at the vicinity of the cold surface descend, and this cycle is maintained, generating convection rolls. The complete resolution of the mass, heat transport, and fluid dynamics conservation equations (Navier–Stokes) shows that there is a critical value of the temperature gradient below which the supply in thermal energy is insufficient to compensate for thermal and viscous dissipation and to set the fluid in motion and that, beyond a certain threshold, the system becomes chaotic. Obviously, if one can hope to counteract the drainage by heating a soap film and generating convection rolls, the risk is to evaporate the water even faster.

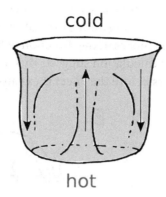

cold

hot

FIGURE 8.16    Rayleigh–Bénard instability in a pan.

## 8.7.2 Cyclone Formation in a Soap Bubble

Hamid Kellay's group at the University of Bordeaux has created a device to form, heat, supply water and visualize a half-bubble of soap. Their device consists of a hollow brass ring with an inlet and an outlet for the circulation of water from the thermostat. This ring has a circular slit filled with soapy water that serves as a water tank. Half-bubbles are blown with a straw. The visualization of the half-bubble is performed in white light through a sheet of tracing paper to ensure scattering of light over a large area.

In the absence of a temperature difference $\Delta T$, colored interference fringes are arranged as parallel layers (Figure 8.17A). This stratification is the signature of continuous drainage. In the presence of a $\Delta T$ slightly above the threshold for convection, convection plumes appear at the equator and rise towards the pole (Figure 8.17B). For high $\Delta T$, drainage layers disappear completely, and disordered convection tends to homogenize the thickness of the soap film (Figure 8.17C).

Hamid Kellay and his colleagues also observed that some plumes generated at the equator further transformed into vortices towards the pole. Remarkably, these vortices within soap bubbles are quite similar to hurricanes visualized on meteorological maps (Figure 8.17E). By changing scale, analyzing and comparing the trajectories, speeds, and lifetimes of vortexes in their bubbles and in tropical cyclones, Bordeaux scientists have shown that there is an exact analogy between the two phenomena. Both systems can be considered pseudo-bidimensional because the ratio of the thickness of the Earth's atmosphere to the size of the Earth (100 km/12,700 km) is much lower than one, and in particular, both types of cyclones, whose movement is trochoidal, undergo a velocity increase to a maximum value, before observing a decrease and complete disappearance. With the help of this soap bubble model, it was also possible to establish a robust law on cyclone movement, allowing a more reliable prediction. Indeed, by a statistical study of trajectory deviations, characterized by mean square displacement (MSD) $\langle r^2(\tau)\rangle = \left\langle \left[r(t+\tau)-r(t)\right]^2 \right\rangle$ (see Section 2.1), Kellay and his collaborators found that MSD $\sim t^{1.6}$. This means that the motion is not Brownian but over-diffusive. More precisely, these vortex and cyclone trajectories are those known as "Levy flights," which can be schematized by an erratic trajectory whose separation intervals between each displacement point are irregular over time. While the origin of this type of displacement is still being debated, the fact remains that this study makes it possible to study simply and therefore to better understand and anticipate cyclone and hurricane movements.

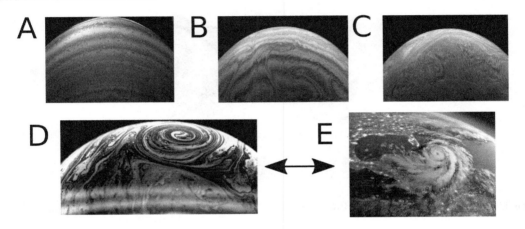

FIGURE 8.17    Image of a half bubble: (A) $T=0$: drainage. (B) $T=11°C$: appearance of convection feathers. (C) $T=40°C$: disordered convection over the entire bubble. (D, E) cyclones in a bubble and in Earth's atmosphere. (A–D: adapted from Fanny Seychelles' thesis; E: copyright Shutterstock.)

## 8.8 SILLY PUTTY

### 8.8.1 Historical Background

Silly Putty is now a toy for children and an anti-stress object for adults. However, its invention originates from research works during World War II aimed at synthesizing artificial rubber. When Japan took the control over most rubber-producing Asian countries, the risk of rubber shortage for the US Army, which needed rubber to produce truck tires, boots, and aircraft parts, led them to trigger intensive research programs for synthesizing rubber. Silly Putty is the success story of an initial failure. Indeed, in two companies, General Electrics and Dow Corning, two chemists, James Wright [1] and Earl Warwick [2], independently mixed silicone oil with boric acid. In 1943, they both patented their invention even though the properties of the obtained putty clearly did not match the expected military requirements. Indeed, the compound exhibits very unusual features: it bounces when thrown against a wall and flows like honey to form a puddle when left unstressed. No use could be found for this bouncing and flowing putty before 1949 when Peter Hodgson, consultant in a toy-store company, decided to produce and commercialize what has become

"Silly Putty" after thinking that the compound could be appealing to both children and parents (Figure 8.18).

FIGURE 8.18   The original "Silly Putty" commercialized in the 1950s. (Copyright Shutterstock.)

## 8.8.2 Phenomenological Model

Silly Putty is a well-known example of a viscoelastic material, and its properties can be sensed qualitatively through the above-mentioned observations. To provide a quantitative description, phenomenological rheological models are often used. The simplest models consist of combining springs with dashpots (Figure 8.19). A "Hookean" spring accounts for the elastic response: $\sigma = k\varepsilon$, with $\sigma$ the stress, $\varepsilon$ the strain, and $k$ the spring constant. A "Newtonian" dashpot allows the modeling of the viscous dissipation: $\sigma = \eta\dot{\varepsilon}$, with $\eta$ the viscosity.

FIGURE 8.19   (A) Kelvin–Voigt model. (B) Maxwell model. (C) Standard linear solid (or Zener) model.

The Kelvin–Voigt model consists of a spring in parallel with a dashpot and is described by the constitutive equation: $(1/k)\sigma = \varepsilon + (\eta/k)\dot{\varepsilon}$. The deformation is restricted by the response of the spring. This model is adequate to describe a viscoelastic solid, i.e. a material that mainly behaves like an elastic solid but which contains some viscous dissipation.

The Maxwell model consists of a spring in series with a dashpot and is described by the constitutive equation: $\eta\dot{\varepsilon} = \sigma + (\eta/k)\dot{\sigma}$. The deformation of the spring will be finite while the dashpot will keep on deforming provided that the load is maintained. This model is more appropriate to describe a viscoelastic liquid.

Note that $\eta/k = \tau$ has the dimensions of time and represents the characteristic relaxation time of transient strain or stress application.

Silly Putty has an intermediate behavior. As shown in Figure 8.19, one needs to add a spring $K$ in parallel with a $(k,\eta)$ Maxwell model in order to account for both creep and residual rubber stiffness after stress relaxation. This corresponds to the so-called Zener of the standard linear solid model. By considering that the overall stress is $\sigma = K\varepsilon + \sigma_{\text{Maxwell}}$ and $\varepsilon = \varepsilon_{\text{spring } k} + \varepsilon_{\text{dashpot } \eta}$, we find the following constitutive equation:

$$\dot{\sigma} + \frac{1}{\tau}\sigma = (k+K)\dot{\varepsilon} + \frac{K}{\tau}\varepsilon \tag{8.1}$$

To perform a frequency analysis, we solve Equation 8.1 in the Fourier transform space with the following notations: $\tilde{\sigma}(\omega) = FT(\sigma(t))$, $\tilde{\varepsilon}(\omega) = FT(\varepsilon(t))$, and thus $j\tilde{\sigma}(\omega) = FT(\dot{\sigma}(t))$, $j\tilde{\varepsilon}(\omega) = FT(\dot{\varepsilon}(t))$. We obtain: $\dfrac{\tilde{\sigma}(\omega)}{\tilde{\varepsilon}(\omega)} = \dfrac{K + j(k+K)\omega\tau}{1 + j\omega\tau}$, which defines a complex modulus $\tilde{G}(\omega) = (\tilde{\sigma}(\omega)/\tilde{\varepsilon}(\omega))$. The real part is called the storage or elastic modulus, $\Re[\tilde{G}(\omega)] = G'(\omega)$, and the imaginary part represents the loss or viscous modulus, $\Im[\tilde{G}(\omega)] = G''(\omega)$.

$G'(\omega) = ((K + (k+K)\omega^2\tau^2)/(1 + \omega^2\tau^2))$ and $G''(\omega) = (k\omega\tau/1 + \omega^2\tau^2)$.

These two moduli are plotted in Figure 8.20.

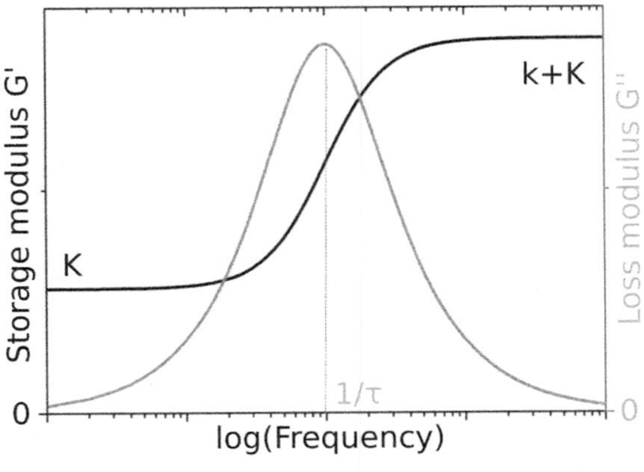

FIGURE 8.20   Variation of the storage and loss moduli, G' and G" as a function of frequency ω.

The elasticity of Silly Putty is maximal at high frequencies, i.e. at short times. Typically, the duration of the impact is a fraction of a second, corresponding to frequencies > 10 Hz. The curve for the viscosity has a bell shape, meaning that the viscosity reaches low values at high and low frequencies. The peak of maximum viscosity corresponds to the time scale $t$, which is typically on the order of 0.1 s for Silly Putty [3].

### 8.8.3 Microscopic Explanation

The main components of Silly Putty are polydimethyl-siloxane (PDMS, or silicone-based polymer) and boric acid (or borax). The PDMS are linear chains that are end-hydrolyzed by boric acid $B(OH)_3$. The trivalent nature of boron allows the linking of two PDMS chains end-to-end but also the crosslinking and formation of a three-dimensional network. When short duration solicitations (such as impact during bouncing) are applied, the Silly Putty behaves like an elastomer (or rubber) and deforms elastically: all B–O–Si bonds remain intact. However, if long duration solicitations are applied (such as under the effect of gravity at rest), the B–O–Si bonds, which are reversible, may unbind, and the observed flow of the Putty is the macroscopic consequence of PDMS chain reptation.

### References

1. J.G.E. Wright, Patent Date:1944-12-23. US2541851A.
2. R.R. Mcgregor, E. Leathen Warrick, Patent Date:1943-03-30. US2431878A.
3. R. Cross, *Am. J. Phys.*, 80, 870 (2012).

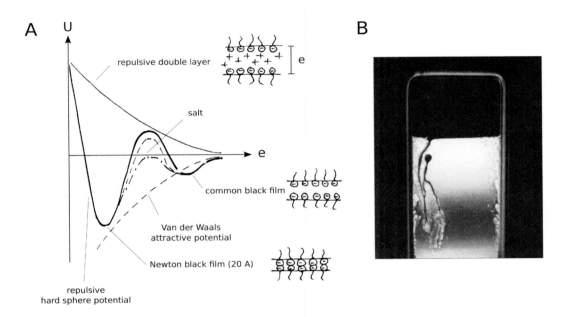

FIGURE 2.26 (A) Classification of soap films from DLVO theory. (B) Draining of a vertical film that becomes a black film. (Photo by K. Mysels.)

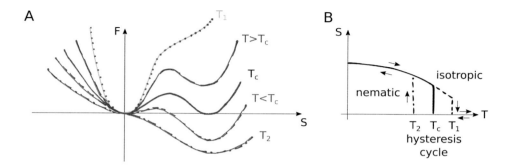

FIGURE 3.11   (A) Free energy isotherms as a function of $S$; (B) Order parameter $S$ as a function of $T$, showing a first-order transition.

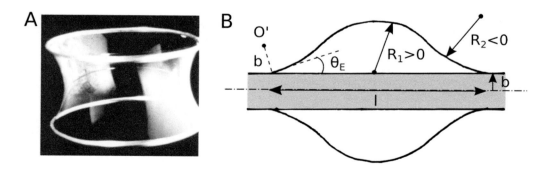

FIGURE 4.11   (A) Soap film between two rings: zero curvature surface. (K. Mysels). (B) Unduloidal shape of a drop on a fiber.

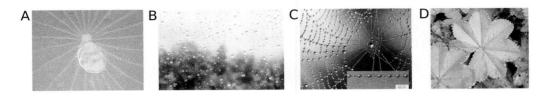

FIGURE 4.20   Examples of wetting in everyday life. (A) Pearl of water on a lotus leaf (photo K. Guevorkian). (B) Drops of water anchored to a window. (Copyright Shutterstock.) (C) Spider web covered with dew (copyright Shutterstock) and zoom on a fiber with its rosary of water drops (F. Vollrath and Edmonds D.T., *Nature* (1989)). (D) Collar of drops on a "smart" sheet, which is hydrophobic to protect and hydrophilic to hydrate. (Photo FBW.)

FIGURE 4.39   Aquaplaning sketched by P.-G. de Gennes. (FBW private collection.)

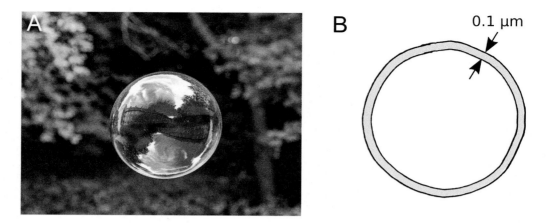

FIGURE 6.13 (A) Soap bubble. (Copyright Shutterstock.) (B) Its membrane has a thickness of about 0.1 μm.

FIGURE 6.15 Vertical soap film obtained by drainage, with a thickness $e(z)$ varying along the vertical axis. At the top the film is black.

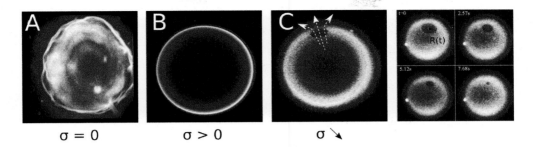

FIGURE 6.21 Transient pore: (A) Floppy vesicle, membrane tension (almost) zero. (B) Tense vesicle after light irradiation. (C) Opening of a pore: $R(t)$ increases. Sequence of four images: pore closure: $R(t)$ decreases. The time is indicated in seconds. (Extracted from [1].)

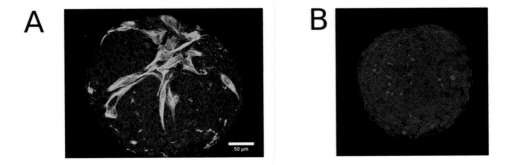

FIGURE 8.6 Images of multicellular spheroids obtained by optical microscopy. Composition: (A) tumor (red) and stromal (green) cells. (Courtesy of Fabien Betillot.) (B) neurons (nuclei in blue, axons in magenta). (Courtesy of Kévin Alessandri.) The diameter of the spheroids is about 300 μm.

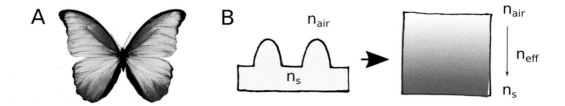

FIGURE 9.2   Anti-reflection graded index films: (A) butterfly at night. (Copyright Shutterstock.) (B) surface with nanometric protuberances to mimic an index gradient.

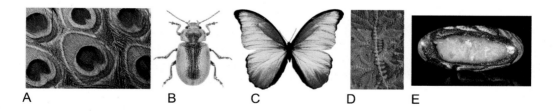

FIGURE 9.11   Examples of photonic structures in nature: (A) A peacock feather; (B) The beetle *Chrysochroa fulgidissima*; (C) The butterfly *Morpho rhetenor*; (D) Young leaves of *Diplazium tomentosum*; (E) An iridescent opal. (A, B, C, E: copyright Shutterstock; D: extracted from [1].)

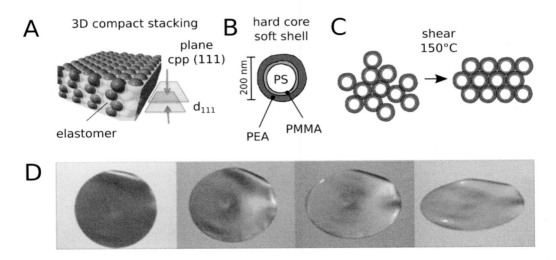

FIGURE 9.12 (A) Polymer opal. (B) Core–shell structure of particles. (Adapted from [2].) (C) Method of preparation of photonic crystal by heating and shearing. (Adapted from [3].) (D) Layers of polymer opal seen from different angles.

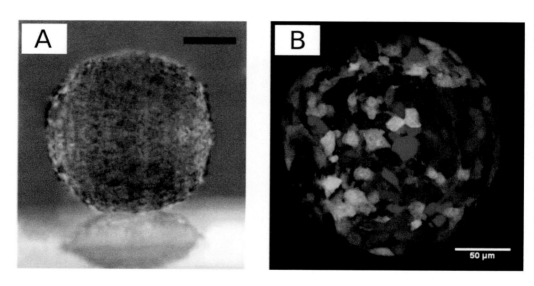

FIGURE 10.26 A multicellular spheroid seen by (A) phase contrast microscopy and (B) confocal fluorescence microscopy. (Courtesy of S. Douezan (A); F. Bertillot (B).)

# CHAPTER 9

# Soft Matter in Technology

## 9.1 ANTI-REFLECTIVE FILMS

### 9.1.1 Katharine Bodgett's Discovery

Katharine Blodgett (Figure 9.1A), whose name is associated with Irvin Langmuir's name for Langmuir–Blodgett (L–B) films, which are ordered molecular layers of amphiphilic molecules adsorbed on a solid surface (Section 6.2), had a peculiar life. Born in 1898 in New York state, she met Langmuir during a visit to the General Electric laboratories where her father worked. Then she went to Great Britain to do her PhD thesis with Ernest Rutherford (Nobel Prize in Chemistry in 1906). In 1926, she became the first woman to graduate with a doctorate in physics from Cambridge University. She then returned to the United States to work at General Electric as Langmuir's assistant on organic monomolecular films (~3 nm) deposited on liquid. This work would allow Langmuir to be awarded the Nobel Prize in Chemistry in 1932. After forming calcium stearate films [$(C_{17}H_{35}COO)_2Ca$], Blodgett realized that, upon dipping a glass slide into the water and slowly removing it when the monomolecular film was formed on the water surface, the glass slide came out dry. This led Blodgett to propose that the hydrophilic heads $(COO)_2^-$ $Ca^{2+}$ attach to the glass by exposing the hydrophobic chain to air, which causes the water film to be withdrawn. From there, Langmuir and Blodgett were able to produce thicker films by successive soakings. Katharine Blodgett then understood the practical importance of these L–B layers in producing anti-reflective films on glass. General Electric patented the idea under the cryptic title "Film Structure and Method of Preparation" (U.S. patent 2,220,660 -16 March 1938), and Blodgett published in 1939 in *Physical Review* an article entitled "Using Interference to Turn Off the Reflection of Glass Light." In particular, she showed an instrument with a glass dial illuminated by a stream of light, only half of which was treated with a L–B coating (Figure 9.1B). On the untreated part there is strong reflection and a dazzling effect which prevent the indications from being read, and this is not the case for the part treated with anti-reflection coating.

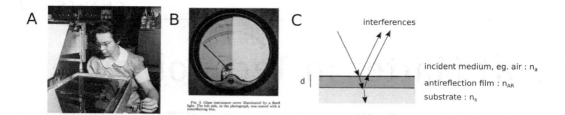

FIGURE 9.1 (A) Photograph showing Katharine Blodgett at work in front of a Langmuir trough. (B) Photograph taken from Katharine Blodgett's article showing the anti-reflective effect. (C) Working principle of anti-reflective treatments.

## 9.1.2 Working Principle of Anti-Reflective Coating

Although glass is transparent, Fresnel's law indicates that in normal incidence the reflectivity of glass (refractive index $n_v = 1.5$) in air ($n_a = 1$) is

$$R = \left( \frac{n_a - n_c}{n_a + n_c} \right)^2 \approx 4\%.$$

To reduce this reflection, the principle consists of depositing a film (of refractive index $n_{AR}$) on the surface of the glass, thus creating two interfaces, respectively air–LB and LB–glass, which produce two reflected light beams that interfere. For a monochromatic light beam (wavelength $\lambda$), the interference will be destructive if the optical path difference between the two reflections is half of an integer times of $\lambda$ (Figure 9.1C). This corresponds to a film thickness of a multiple of ¼ of wavelength inside the film, i.e. $\lambda/4n_{AR}$. For the interference to be completely destructive, the amplitudes of the two reflections must be equal, which is obtained if $n_{AR}$ is the geometric mean of the air and substrate indices $\left( n_{AR} = \sqrt{n_a \cdot n_v} = 1.22 \right)$. On the other hand, extinction is only valid for one wavelength of light, not for the entire spectrum. In practice, Katharine Blodgett could not adjust the optical index of the L–B films (~1.54 for cadmium arachidate films), and she optimized the number of layers with respect to $\lambda = 550$ nm, close to the center of the visible spectrum and corresponding to the maximum sensitivity for the human eye.

## 9.1.3 Technological Progress and Natural Inspiration

Remarkably, Blodgett's method was used as early as 1939 by the directors of the movie "Gone with the Wind" to treat the lenses of the cameras. However, while all eyeglass lenses, camera lenses, solar cells, and television screens are now covered with anti-reflective coatings, the process has been improved since then. A universally used material is magnesium fluoride ($MgF_2$) whose index ($n = 1.38$) is closer to the theoretical optimal index value and whose vacuum evaporation allows a better adhesion to the glass, thus robustness, than adsorbed L–B films. To extend the anti-reflective properties to all wavelengths and avoid the colored appearance of monolayer-based treatments, the idea is to design multilayer stacks of variable thickness and index. This optimization requires opticians to accurately calculate the reflection coefficients at all interfaces and the amplitudes of the reflected light beams. In this approach, low index media are deposited alternately with high index media. An alternative approach also consists of avoiding sudden changes in the refractive index between a material and its surrounding environment by ideally using a graded index coating, which is based on the progressive adaptation of the refractive index between the incident medium and the material, typically between air and glass. Indeed, Lord Rayleigh, as early as 1879, described the effect of a density gradient (and therefore an index gradient) between two media to "curve" the path of a light beam and thus avoid reflections. This leads to what is known as the "mirage effect" observed on summer roads. However, it is quite a technical challenge to manufacture in a controlled manner a gradient of refractive index. A clever approach is inspired by the camouflage strategy of some nocturnal insects, which have also a high capacity to collect the low level of nocturnal light radiation. Moth butterflies are a perfect example (Figure 9.2A): they develop on the surface of their corneas a hexagonal network of conical or cylindrical nanostructures of a size smaller than the visible wavelengths ($\approx$ 100 nm). These periodic structures have an anti-reflective effect because they generate a refractive index gradient due to the geometry of the nanometric protuberances and allow a progressive decrease in the index from 1.54 (chitin) to $n_a = 1$, as shown in Figure 9.2B. The reproduction of nanostructured surfaces similar to those observed in nature is the subject of active research. The Taiwan group of Professor H.H. Yu has shown that one of the most effective ways to prepare such protuberance networks is to make Langmuir–Blodgett monolayers of plastic nanospheres. Why are latex beads stable on the surface of water, and do they behave like macro-amphiphiles? This question was solved almost 40 years ago by Pawel Pieranski [1]. Polystyrene sulfonate beads deposited on the surface of the water are partially immersed, and their ionizable groups (sodium sulfonate) are dissociated only on this submerged part, which leads to an asymmetric distribution of charges. In the presence of counterions, this gives rise to a double electric layer and thus an effective macroscopic dipole. The repulsion between the aligned dipoles is at the origin of the formation of colloidal networks, similar to the ones found on the surface of the moth butterfly eye.

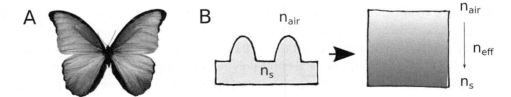

FIGURE 9.2 **(See color insert.)** Anti-reflection graded index films: (A) butterfly at night. (Copyright Shutterstock.) (B) surface with nanometric protuberances to mimic an index gradient.

## References

1. P. Pieranski, *Phys. Rev. Lett.*, 45, 569–572 (1980).
2. W.K. Kuo et al., *ACS Appl. Mater. Interfaces*, 8, 32021–32030 (2016).

## 9.2 ARTIFICIAL MUSCLES BASED ON NEMATIC LIQUID CRYSTALS

### 9.2.1 Generalities on Muscles and Artificial Muscles

Muscles are composed of muscle fibers, which are long polynuclear cylindrical cells. Each fiber itself is made up of bundles of myofibrils composed of sarcomers. Sarcomers, which are the elementary contractile units, are composed of two types of filaments (rich in actin and myosin proteins – Section 10.4) that slide relative to each other and contribute to the lengthening or shortening of the muscle fiber, thus generating muscle contraction.

With the notable exception of reflex motions, muscle contractions result from a conscious effort that comes from the brain. It sends stimuli, in this case electrochemical signals, through the nervous system to motor neurons, which are in direct contact with the muscle fiber through the neuromuscular junction. It is at this synapse that neurotransmitters (acetylcholine) are released and trigger a biochemical cascade allowing the muscle to contract.

This global process, which is the result of a complex and multi-scale design, is now well understood. Of course, currently for applications in regenerative medicine or previously to manufacture robots, developing strategies to manufacture artificial muscles is a very active field of research. Obviously, the specifications have to be adapted to the proposed application. For example, solid actuators (piezoelectric or ferroelectric) that deform under the action of an external field have made possible the production of many robotic and prosthetic devices, often even more effective and responsive than natural muscles, that do not "feel" fatigue and can produce much higher forces. On the other hand, they are heavy, rigid, and unsuitable for fine movements such as those that contribute to facial expressions or, more simply, to grabbing a ball. Many other approaches have been developed, and it would take too long to do a full review of their relative pros and cons.

### 9.2.2 A pH Based Chemical Muscle

We would like to mention here a physicochemical approach proposed by P.-G. de Gennes in 1997 [1]. This was based on an idea of Katchalsky's in 1949, which consists of exploiting the possibility of transforming chemical energy into mechanical energy using inflated hydrogels. In practice, this first realization used polymer chains with carboxylic acid groups (COOH). Upon sodium hydroxide addition ($OH^-$ ions), the acid groups become carboxylate groups ($COO^-$): this stretches the flexible polymer chains by introducing some electrostatic repulsions. The addition of hydrochloric acid ($H^+$ ions) re-protonates the carboxylate groups causing the gel to contract to its initial state. In order to transfer these molecular extensions and contractions to a more macroscopic scale, Katchalsky crosslinked the polymer chains (by using the esterification of polyacrylic acid with polyhydric alcohols). Figure 9.3 shows the principle of operation of this "pH based muscle." The main problem with this system is that $Na^+$ and $Cl^-$ ions are added to each cycle, which eventually screen totally the electrostatic interactions. Another limitation is the speed of the response, which is limited by the diffusion of $OH^-$ or $H^+$ ions into the gel: each extension or relaxation takes a few minutes to establish ... which is far too slow compared to biological muscles !

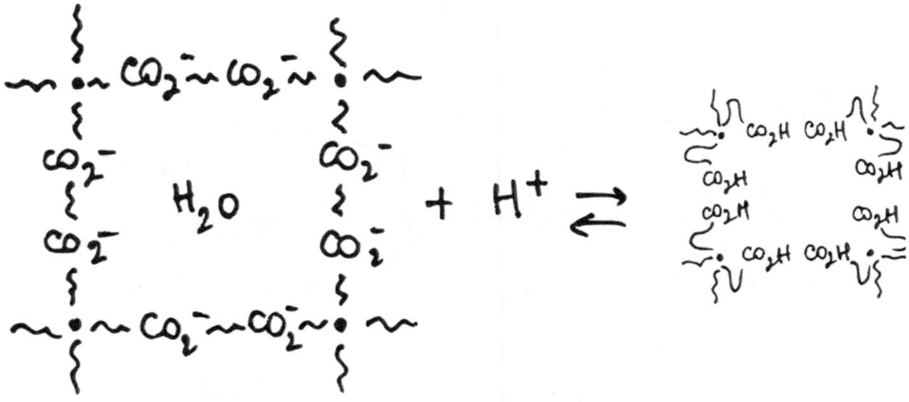

FIGURE 9.3    Principle of a "pH based muscle" as proposed by Katchalsky in 1949.

### 9.2.3  De Gennes' Nematic Muscle

Building on this idea, de Gennes suggested the use of the rapid response of liquid crystals to physical stimuli such as temperature, light, or the electric field. For example, using a thought experiment, a nematic network (Figure 9.4) below its nematic–isotropic transition temperature $T_N$ is elongated. When the temperature is raised above $T_N$, the mesogenic groups become disordered, and the network contracts.

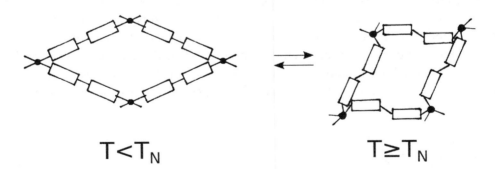

$$T < T_N \qquad\qquad T \geq T_N$$

FIGURE 9.4    Principle of the "nematic muscle."

The effect will even be more significant if the nematic networks are monodomain, i.e. if the mesogens are macroscopically oriented according to the director vector. To promote this, P.-G. de Gennes suggested to chemists to use triblock copolymers with mesogen-free flexible portions (C) for gel crosslinking, alternating with nematic portions (N) (Figure 9.5). The diffusion of heat being faster than that of molecules, this thermo-responsive nematic system is a priori faster to change shape. A heat pulse can be achieved quickly, by laser or by introducing dopants such as carbon nanotubes, but to complete the cycle, it is necessary to cool down, and this process is indeed much longer, hence limiting the system response times.

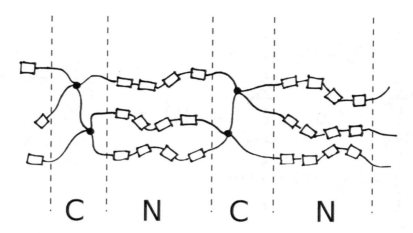

FIGURE 9.5   Possible practical realization of a liquid crystal muscle consisting of rubber portions (C) and nematic portions (N).

## 9.2.4 Fabrication of Semi-Fast Artificial Muscles

M.H. Li and P. Keller, two chemists from the Institut Curie, adapted experimentally this artificial muscle design according to the constraints of synthesis and in particular had to consider another type of stimulus [2]. Rather than using so-called main-chain polymers for which mesogens are inserted into the core of the main chain, they have synthesized side-chain polymers for which mesogens are laterally attached to the polymer chain (Figure 9.6A). The selected mesogenic group contains an azobenzene function, which is known to undergo *cis–trans* isomerization under UV light irradiation. The *trans* isomer, whose shape is rod-like, has mesogenic properties, while the curved *cis* isoform is not mesogenic. Irradiation at 365 nm therefore induces a nematic–isotropic transition of the nematic elastomer (Figure 9.6B). On oriented films, contractions of 15–20% in about ten seconds were obtained.

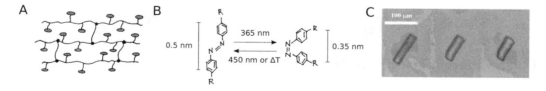

FIGURE 9.6   (A) Side-chain cross-linked nematic polymers. (B) Reactive entity and reversible conformational changes under illumination. (C) Reaction realization (from left to right, contraction under illumination).

With the help of A. Buguin, a physicist specializing in microfabrication, also from the Institut Curie, they produced small pillars 20 μm in diameter and 100 μm in length by crosslinking the molecules under a magnetic field to facilitate orientation. They obtained a contraction of these micro-actuators reaching 35% (Figure 9.6C). On the other hand, reversibility of the system, using either irradiation with visible light or by heating, is much less effective, again limiting the system.

Applications of these artificial liquid crystal "muscles" for everyday life are still far away, and these achievements are probably worth more for their conceptual approach. Nevertheless, this shows that liquid crystals can still have unexpected applications beyond their usual use for display functions in TV sets.

References

1. P.-G. de Gennes, *C. R. Acad. Sci. Sér. II*, 324, 343–348 (1997).
2. M. H. Li, P. Keller, *Philos. Trans. R. Soc. A*, 364, 2763–2777 (2006).

## 9.3 THE MAGIC OF PAINTING

The chemical basis of paint is a polymer resin, to which pigments and other additives are added. To form a homogeneous film of paint on a wall, it is important to ensure that the paint is viscous enough to avoid it flowing down once spread on the wall. But we forget that if the polymer contained in a bucket of paint and present as small droplets of polymer melts was put in solution, the paint would be so viscous that we would not be able to mix it. The secret of painting lies in the use of latex polymer.

### 9.3.1 Generalities on Latex Particles

A latex is an aqueous colloidal suspension of polymer particles. The term latex comes from the Latin word meaning liquor and was adopted in the 17th century to designate the juice of certain plants such as rubber trees, whose sap is natural latex. These latexes were used by Mayans and Aztecs as waterproofing agents for clothing and for making rudimentary boots. During the Second World War, the German natural rubber industry invented a new synthetic process: emulsion polymerization. The emulsification of a mixture of (ionic) surfactants, water, and monomers allows the insoluble monomers to form droplets. The addition of polymerization initiators triggers polymerization and thus the nucleation of polymer droplets. These grow gradually by diffusion of the monomers, until the monomer is exhausted.

In the preface to a reference book on latex by J.-C. Daniel and C. Pichot [1], P.-G. de Gennes writes:

*These synthetic latexes have shown rare qualities:*

- *The particle/water mixture is not very viscous (much less than a simple polymer solution at the same concentration) – so it is easy to handle.*

- *By spreading and drying, polymer films are formed where the particles are welded together - thanks to fairly fine chemical adjustments.*

- *By mixing in the initial drops a "hard" monomer (styrene) and a "soft" monomer (butadiene), the mechanical properties of the final product can be adjusted.*

- *The latex carrier liquid is water: there is no toxic solvent to remove during use.*

Latexes have many applications. They are mainly used as binding agents in painting and paper coating. While the world production of polymers is 200 million tons, that of latex is about 10%, which represents a considerable market.

### 9.3.2 Viscosity of Latex

To illustrate the first feature of latex highlighted by P.-G. de Gennes, namely a high polymer content for a low viscosity, we can take the example of Sterocoll, sold by BASF as a thickening agent, which is an aqueous acid polymer dispersion containing polymer particles rich in carboxylic acid groups. The polymer dispersion can be dilute with water up to a polymer concentration of about 5%. It is still a white liquid that is only slightly more viscous than water. By adding a small amount of a sodium hydroxide solution, the pH is increased, and the carboxylic acid groups are transformed into negatively charged carboxylates. This makes the polymer particles soluble in water, and a polymer solution is formed. At the same time, the viscosity increases significantly. After a few minutes, a gel is obtained.

### 9.3.3 Formation of a Film of Paint

There are three stages in the formation of latex films by drying (Figure 9.7):

- Water evaporation causes the particles to concentrate until they pile up. Like oranges on a fruit merchant's stall, particles can occupy about 64% of the volume (for random stacking) or 74% of the volume (for compact stacking). The rest is water.

- As the drying process goes on, the particle assembly is compressed. If the particles are non-deformable (Young's modulus > $10^7$ Pa), the process stops there. The film, with holes, is permeable to air, but it is fragile. If some "soft" monomer has been introduced (according to the term used by P.-G. de Gennes), the particles are deformed and take a polyhedral or honeycomb shape. They remain individualized. The film has become more cohesive and continues to "breathe," which is an advantage for building paints. A surface treatment of the latex allows the process to be stopped at this stage. Before moving on to the third stage of drying, let's look at the origin of the forces that lead to latex deformations (Figure 9.8). When contact is made, the water film deepens between the two particles. The surface pressure at the level of the polymer film is given by [1]: $P = \gamma\left(\left(1/r_1\right) - \left(1/r_2\right) + \left(2/r\right)\right)$ with $\gamma$ the polymer–water interfacial tension. We see that when $r_1$ decreases, $P$ increases and can reach $10^7$ Pa for radii of curvature as small as 10 nm and an interfacial tension of about 0.1 J/m$^2$.

- Finally, if the particles are deformable, the polymer chains will diffuse from one particle to its neighbors through the interfaces. The interfaces fade, the particles become indistinguishable, and the film is homogeneous. This process corresponds to the coalescence of latexes.

These film-forming mechanisms are still being investigated from a fundamental point of view, particularly in the context of ageing and cracking formation over time. For example, observations have shown that thick layers crack more often than thin layers of paint, due to mechanical stresses accumulated throughout the film during drying.

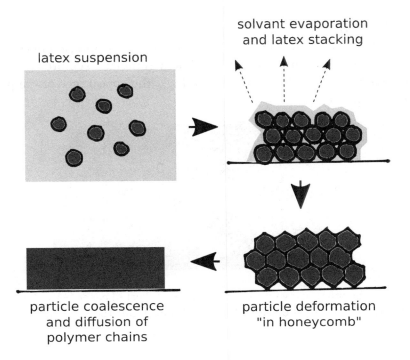

FIGURE 9.7   Schematic representation of the different steps in the process of forming a film from latex.

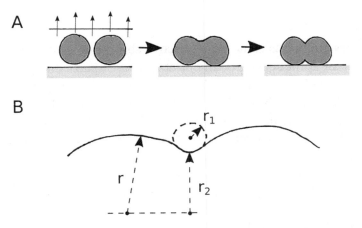

FIGURE 9.8   (A) Focus on the drying step of two latex particles. During water evaporation, the interface "deepens." (B) Definition of the notations used for the different radii of curvature.

Moreover, it is remarkable that by controlling the surface treatment of latexes (i.e. the type of surfactant used to stabilize them), it is possible to produce paints that allow water to pass through or others that are waterproof.

### 9.3.4 Paper Coating

Another important application of latex is paper coating. Depositing a thin layer of latex on paper fibers transforms the fibrous structure into a granular structure (Figure 9.9), with better printability (and therefore better rendering). The appearance of the paper (gloss, whiteness) is also improved.

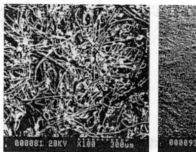

FIGURE 9.9   Electron microscopy images of an uncoated sheet of paper (A) and a coated sheet of paper (B), i.e. coated with a layer of latex. (Extracted from "Industrial Latex Applications" by Dow Emulsion Polymers.)

### Reference

1. J.C. Daniel, C. Pichot, *Les latex synthétiques*. Lavoisier/Tec et Doc, 2006.

## 9.4  IRIDESCENT CLOTHES

Most of the colors around us come from selective absorption caused by pigments incorporated into bulk or at the surface of an object. However, particularly intense and bright colors result from the interaction of light with micro- and nanostructures, which cause interference (Section 9.1) or diffraction. Animals and plants use these so-called photonic structures (i.e. regular structures with a periodicity in the order of the wavelength of visible light) to produce fascinating optical effects. These structural colors can be either iridescent if they change according to the angle of view or non-iridescent in the opposite case. This has become a source of inspiration for technological applications such as the manufacture of iridescent clothing that changes color as it is stretched or wrinkled.

### 9.4.1  What Is a Photonic Crystal?

A photonic crystal is a material whose refractive index varies periodically (on a length scale comparable to the wavelengths of light). Multilayer stacks (or Bragg mirrors) are often referred to as one-dimensional photonic crystals with a periodicity perpendicular to the planes of the layers.

By reasoning on the interferences created at each interface between two media of optical index $n_a$ and $n_b$, of thickness $d_a$ and $d_b$, and by defining the refraction angles $\theta_a$ and $\theta_b$ (Figure 9.10) and the angle $\theta_0$ of incidence from the air of index $n_0$, we show that there is a reflection peak at a wavelength $\lambda_{max}$ (for the angle $\theta_0$) expressed by :

$$m\lambda_{\max} = 2\left(d_a\sqrt{n_a^2 - n_0^2 \sin^2\theta_0} + d_b\sqrt{n_b^2 - n_0^2 \sin^2\theta_0}\right), \tag{9.1}$$

with $m$ an integer.

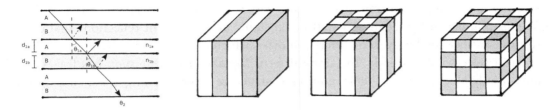

FIGURE 9.10  Multilayer film and notations used for interference calculation. 1D, 2D, and 3D photonic crystals.

When the object is illuminated with white light, we mostly see the specific color corresponding to $\lambda_{\max}$. If the bottom layer of the stack is dark, it absorbs all the light that has not yet been reflected, making the reflected color very bright. When the base layer becomes lighter, it reflects more light at all frequencies, and this light is then transmitted to the upper layers, giving a less bright effect or a silvery or golden appearance.

A more complete calculation (and one applicable to 2D and 3D photonic crystals) is based on a plane wave expansion from the Maxwell equations to derive a band diagram. Propagation equations yield the evolution of the frequency $\omega$ as a function of the wave vector $\vec{k}$ (perpendicular to the wave front and standard $2\pi/\lambda$) and show the appearance of forbidden frequency (and therefore energy) bands. A range of frequencies forbidden for propagation corresponds to reflected waves, as shown in the case of the 1D system.

### 9.4.2 Photonic Structures in Nature

There are many species of flora and fauna that use photonic structures to create structural colors [1] and iridescence for camouflage, sexual selection, or pollinator attraction (Figure 9.11).

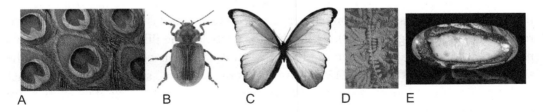

A        B        C        D        E

FIGURE 9.11  **(See color insert.)** Examples of photonic structures in nature: (A) A peacock feather; (B) The beetle *Chrysochroa fulgidissima*; (C) The butterfly *Morpho rhetenor*; (D) Young leaves of *Diplazium tomentosum*; (E) An iridescent opal. (A, B, C, E: copyright Shutterstock; D: extracted from [1].)

As early as 1665, Robert Hooke, in his book *Micrographia: or Some physiological descriptions of minute bodies made by magnifying glasses*, reported observations of a peacock feather changing color according to the angle from which it is viewed.

A more classical and much studied example is the butterfly wing. For example, *Morpho rhetenor* has a wing that is micro-structured with Christmas tree scales. The *Chrysochroa fulgidissima* beetle (or Japanese jewel) has metallic and iridescent reflections that come from a stratification by layers of different indices, which causes light interference and high spectral selectivity. Similarly, the tropical fern, *Diplazium tomentosum*, has leaves that are initially blue because of their multi-layer structure before turning green later during growth.

In the mineral field, we also find iridescent opals. The most famous opal-AG is composed of silica microspheres that are formed by slow hydrolysis reactions and organized in successive layers.

In addition, some creatures are able to reversibly change their structural colors according to their environment. This is the case for the beetle *Tmesisternus isabellae*, which has golden reflections in a dry state and turns red when it becomes wet. A physicochemical characterization of the multilayer structure of its scales reveals a low contact angle, thus an ability for water to adsorb and then infiltrate. In doing so, the refractive index of the photonic structure is modified, and the band of reflected wavelengths is as well.

### 9.4.3 Iridescent Clothing

All these natural observations inspired chemists who wanted to create iridescent materials by inclusion of photonic structures. Composite opals, and more particularly polymers, have recently been developed ([2] and [3]). Figure 9.12A shows a drawing of a synthetic polymer opal, which consists of a 3D crystal of particles in an elastomeric matrix. The technical difficulty comes from the need to produce a crystalline structure, which has properties similar to a Bragg mirror, in order to generate light interference. Jeremy Baumberg's group in Cambridge, UK has chosen an original approach by synthesizing sub-micrometric particles of the core–shell type: the core made of PS (polystyrene) is hard, and the shell made of poly(ethyl acrylate) (PEA) is soft. Here, the soft/hard distinction comes from the fact that the PEA has a glass transition temperature around $0°C$ and is therefore viscoelastic at room temperature (Figure 9.12B). By heating the particles, the PEA shells form a viscous matrix that fills the gaps between the PS cores. Stratification is obtained by shearing the sample or extruding it (Figure 9.12C). This results in cubic-type structures with compact packing (ccp). The colors range from red to green depending on the angle at which they are observed (Figure 9.12D).

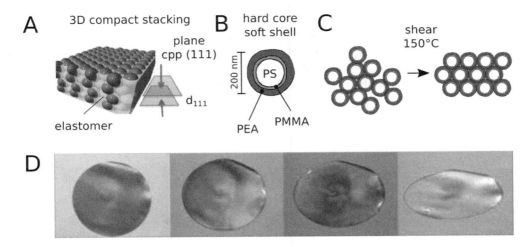

FIGURE 9.12 **(See color insert.)** (A) Polymer opal. (B) Core–shell structure of particles. (Adapted from [2].) (C) Method of preparation of photonic crystal by heating and shearing. (Adapted from [3].) (D) Layers of polymer opal seen from different angles.

After crosslinking, these opals become elastomers and can be shaped into sheets or fibers with very bright structural colors. More importantly, their elastomeric properties make them stretchable. Under stretching, the distance between particles in the same plane increases, but the distance between the different layers of the arrangement decreases (due to volume conservation). According to Equation 9.1, as the distance between layers decreases, the wavelength decreases, corresponding to a transition from blue to red, through green and yellow.

A valorization of this invention consisted of making iridescent clothes with bright colors which changed with the movements of the body (Figure 9.13). An English fashion designer used such clothes during a haute couture fashion show in Paris. When will we see such clothes in our daily lives?

FIGURE 9.13  Examples of elastomeric opal clothing.

## References

1. J. Sun, B. Bhushan, J. Tong, *RSC Adv.,* 3, 14862–14889 (2013).
2. C.E. Finalyson C. Goddart, E. Papachristodoulou, D.R.E. Snoswell, A. Kontogeorgos, P. Spahn, G.P. Hellmann, O. Hess, J.J. Baumberg, *Optics Express*, 19, 3144–3154 (2011).
3. C.E. Finalyson, J.J. Baumberg, *Polym. Int.*, 62, 1403–1407 (2013).

# CHAPTER 10

# Soft Matter in Biology

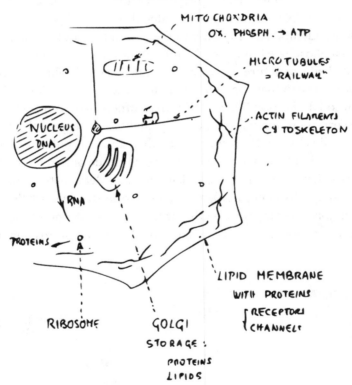

EUCARYOTIC CELL

Here is a drawing of a cell by P.-G. de Gennes (from FBW private archives). Many cell constituents, including DNA inside the nucleus, the lipid membrane, and the cytoskeletal filaments, can be described using soft matter physics concepts. The following sections will exemplify this approach.

## 10.1 ELASTICITY AND COMPACTION OF DNA

### 10.1.1 How to Measure the Elasticity of Single Polymer Chains?

Polymers are a class of soft matter materials. Like liquid crystals or foams, they are characterized by a huge response to weak perturbations. These responses can be observed on macroscopic systems such as a British jelly or a rubber band which deform much more than an iron block or a wood rod under traction. But what happens at the molecular level?

The ideal thought experiment is to grab a polymer chain from both ends and move them away from each other (Figure 10.1A). Simultaneous measurement of the applied force as a function of elongation should allow us to determine the elasticity of the polymer chain. In practice however, it is difficult to synthesize polymers of controlled and high enough molecular mass so that they become visible to the naked eye and thus can be handled. In contrast, DNA, which is condensed in each human chromosome, has an end-to-end length of a few centimeters, and most bacteriophage DNA molecules (which infect bacteria) are several tens of microns. DNA is therefore considered as a model polymer for biophysicists. The bases of DNA, adenine (A), thymine (T), guanine (G), and cytosine (C), associated with a sugar and a phosphate group are the nucleotides that form the monomers of DNA. Its special double-helix structure makes it a so-called semi-flexible polymer (see Section 7.2). Its behavior is close to that of an ideal polymer. The configuration of DNA in water is that of a random coil. In Section 7.2, we derive the stiffness of this entropic spring. Without referring to the corresponding formula, we may qualitatively expect that the force needed to "unwind" the coil is low, in the same way as when a yarn is pulled out of a ball of wool (provided there is no knot). What is the order of magnitude of this force? Additionally, as for most materials that exhibit linear elasticity at small deformations (i.e. the measured force is proportional to its elongation), a deviation from this linear behavior is expected at larger deformations, accompanied with hardening or softening of the material. What about DNA? How to measure the force–extension relationship of this polymer coil?

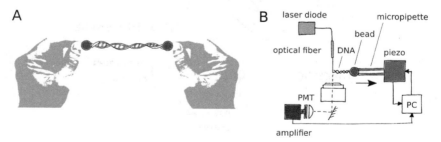

FIGURE 10.1   A) Thought experiment aimed at measuring the elastic properties of a DNA molecule. B) Experimental setup designed to measure forces by using a microfiber. (Adapted from [2].)

In the mid-1990s, the groups of Carlos Bustamente at the University of Oregon [1] and the one of Didier Chatenay at the Institut Curie [2] carried out the experiment proposed in Figure 10.1A. Although the experimental setup was slightly different, the working principle remained the same. We describe here the experiment of the French group. The selected

DNA is that of the λ phage (a virus) which has 48,502 base pairs and a total length of 16.4 µm. To see it under a microscope, one uses fluorescent dyes (named YOYO, TOTO, etc.) which are intercalated between base pairs of the double helix. But the experiment itself was carried out in "blind" conditions, exploiting the fact that the ends of DNA molecules are chemically different from the other groups and therefore can react specifically with certain molecules. These have been functionalized, on one end by a steroid, digoxigenin (Dig), and on the other end by a vitamin, biotin (Biot). Biotin has a high affinity for a protein (found in egg white), avidin (Av), and digoxigenin has a specific antibody, anti-digoxigenin (anti-Dig). As a consequence, when a micro-bead covered with avidin and an optical fiber coated with antidigoxygenin are dipped in a solution containing these functionalized DNA molecules, the extremities of the DNA eventually anchor on one side to the bead and the other to the fiber under the action of thermal agitation. The fiber and the bead thus serve as handles to stretch the DNA molecule. To measure the force, the fiber is used as a dynamometer or force sensor. The displacement of the bead, connected to the fiber via the DNA linker, causes a deflection of the fiber which behaves like a beam. To gain a better force sensitivity, the method was refined by etching up the fiber to a diameter of about 10 µm and by calibrating to determine its rigidity. Then, under a microscope, the distance between the surface of the fiber and that of the bead, which corresponds to the length of the DNA molecule, was measured simultaneously with the deflection of the beam as the bead attached to a micromanipulator was moved away. The scheme of the experiment is sketched in Figure 10.1B.

## 10.1.2 The Elastic Properties of DNA

The measured forces range from 1 pN to 100 pN. The DNA molecule first elongates under weak force. This corresponds to the entropic regime mentioned above. As we approach the contour length of the DNA, the force required to lengthen the DNA further increases, because we begin to stretch the atomic bonds. The worm-like chain model, which is classical in polymer theory, allows us to account for these two regimes. On the other hand, the surprise came from what happened beyond a threshold of about 70 pN. At this stage, the double strand of DNA lengthens abruptly by 60% (Figure 10.2A) without requiring any extra force. Finally, for larger elongations, we find an elastic behavior similar to that observed initially, and the process ends with a break (or detachment of one of the two ends of the DNA molecule). This region of huge elongation at constant force is a plateau (called overstretching) that is reminiscent of first-order phase transitions, such as water boiling: when water is heated around 100°C, it begins to evaporate, and there is coexistence of vapor and liquid water. The temperature remains at 100°C until all liquid water is evaporated, and the additionally supplied energy is used to heat the steam. Here, we have a structural transition of the DNA molecule. Numerical simulations [3] then suggested that DNA overstretching is associated with double-helix unwinding (Figure 10.2B) and separation of base pairs along the molecule axis. The bases are no longer stacked in a helical configuration but almost perpendicular to the axis of the phosphate skeleton, in a "ladder" configuration.

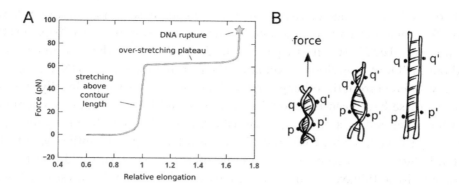

FIGURE 10.2 (A) Force-extension curve for a double-stranded DNA molecule showing the over-stretching transition. (B) Schematic representation of DNA conformations during stretching. Points p, p′, q, and q′ allow the setting of references along the double helix.

At first sight, one could think that this study only shows an exotic DNA structure, generated by the application of high (on this scale) external forces. Actually, this work initiated many further studies in biophysics. Numerous DNA-interacting proteins such as the RecA recombinase protein or the RNA polymerase that performs DNA transcription can only function when DNA is (locally) in this super-stretched configuration.

### 10.1.3 The Puzzle of DNA Accommodation in the Nucleus: The Role of Histones

Besides its singular elastic properties, DNA is also a polyelectrolyte, i.e. a highly charged polymer, since all phosphate groups along its backbone are negatively charged. Biological media are generally salty, with $K^+$, $Cl^-$, $Ca^{2+}$ ions at concentrations greater than 1 mM, which implies that the Debye screening distance is typically between 1 and 10 nm (Section 2.4). As a consequence, electrostatic interactions are usually insignificant at the cellular or tissue level. However, they are crucial within the nucleus. In particular, we will see here how DNA compaction and key steps in cell life like the transcription process require fine control of DNA–protein electrostatic interactions.

The high linear charge density along the DNA primarily affects the "rigidity" of the DNA molecule. In addition to the winding of the two strands into a double helix conformation, and even though the electrostatic screening length is nanometric, the proximity of identical charges disfavors folding and therefore contributes to increasing the length of persistence, which is $l_p = 50$ nm, i.e. about 150 base pairs, for double-stranded DNA. Since the human genome contains approximately $3 \times 10^9$ base pairs, the overall length of the DNA polymer obtained by bridging all chromosomes corresponds to $N = 3 \times 10^9 / 150 = 2 \times 10^7$ "monomeric units" of a "virtual" polymer chain (of segment length $l_p$). According to Section 7.2, the size of the corresponding polymer coil, in an ideal chain model, is $R_0 = N^{1/2} l_p = 250$ μm. This may seem to be small compared to the 2 m contour length of the genome. But how to explain that this genome is accommodated in a nucleus whose size is of the order of 5 μm for mammalian cells?

Surprisingly, the DNA compaction required to accommodate this 250 μm in radius coil into 5 μm in radius nucleus is achieved by adding proteins, called histones. Addition of material is a priori counterintuitive as a route to size reduction. Histones are octameric proteins that have the shape of a small barrel (7 nm in diameter and 6 nm in length) and are highly positively charged (+146 e). The negatively charged DNA molecules bind to histones via coulombic interactions. Minimization of energy (balance between electrostatic gain due to released counter-ions and curvature energy penalty) leads to a mixed DNA–histone structure, called a nucleosome, in which the DNA makes 1.7 turns around each octamer histone. 140 DNA bp are in direct interaction and 60 bp serve as linkers between two nucleosomes. On average, a nucleosome thus consists of one histone octamer and 200 DNA bp. For the entire genome to be compacted into nucleosomes, we therefore need $3 \times 10^9/200 \approx 10^7$ histones.

The nucleosome, shown in Figure 10.3, is thus roughly a cylinder of 6 nm in length and $7 + 2 \times 2 = 11$ nm in diameter, which means that it occupies a volume $\pi d^2 h/4 \approx 570$ nm$^3$. The volume occupied by $10^7$ nucleosomes is therefore 0.05 μm$^3$, which is 100 times less than the volume of the nucleus! The addition of 10 million octamers of histones thus allows the compaction of DNA into the nucleus. Note that this structure called "beads-on-wire" is only the first step in the compaction of DNA, which continues with a chromatin fiber made of nucleosomes, then with extended and condensed chromosome forms.

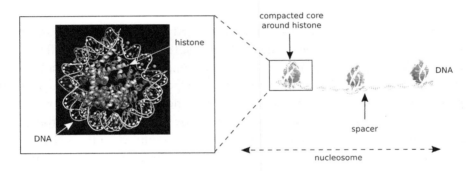

FIGURE 10.3    Structure of a nucleosome. (Copyright Shutterstock.)

While this compaction of DNA around histones mediated by strong electrostatic interactions is beneficial to accommodate the DNA in the nucleus, this organization based on intimate DNA–protein binding a priori generates another issue related to the accessibility of the DNA bases during transcription. This process is the first step that allows the use of the genetic information carried by the succession of DNA bases and the transformation of it into proteins that will accomplish multiple functions within the cell. The synthesis of messenger RNA, which is an intermediate between DNA and proteins, involves the transcription of the genome: the RNA polymerase is a protein complex that has the function of "transcribing" one of the two strands of DNA into an RNA sequence (single strand). To do this, the double helix must be partially denatured, i.e. the hydrogen bonds between DNA bases are locally broken, permitting RNA polymerase binding and base accessibility (Figure 10.4). Steve Block and his collaborators [4] performed a conceptually simple in vitro experiment and showed that the RNA polymerase exerts a force of about 5 pN to move along a DNA molecule while opening it and polymerizing the RNA sequence in the presence of nucleotides. However, other in vitro experiments revealed that the electrostatic interaction within a nucleosome between DNA and histones required a force of at least 20 pN [5] to disrupt the structure and unwrap DNA. In other words, nucleosomes act as a barrier to transcription, preventing RNA polymerase from sliding along DNA and synthesizing RNA. How to explain that, in vivo, RNA polymerase succeeds in performing its transcription work?

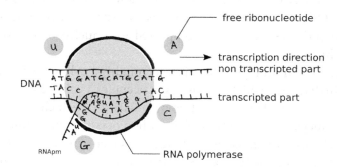

FIGURE 10.4   Transcription of DNA by RNA polymerase.

Carlos Bustamente and his colleagues [6] proposed and showed that transcription is permitted and regulated by nucleosome fluctuations, that is, by a kind of "breathing" of nucleosomes. The experiment consists of attaching a single nucleosome, via a DNA spacer, between two beads trapped in a double optical trap. Upon the addition of nucleotides, the RNA polymerase slides along the DNA. In the presence of salt (KCl) in low concentration, the charges of opposite sign carried by the DNA and the histones strongly attract. The interaction energy is, in absolute value, high. As a result, when the RNA polymerase encounters a base of the histone-bound DNA, it cannot proceed any longer and remains stalled at this stage of transcription. On the other hand, for higher salt concentrations, electrostatic interactions are more labile. DNA and histone are sometimes in intimate contact, sometimes locally detached. This fluctuating barrier therefore allows the RNA polymerase to be stalled only in a transient fashion. The frequency and duration of these pauses during transcription are thus governed by the ionic strength of the solution. The RNA polymerase does not need to develop significant forces to unwind the DNA wrapped around the histones; it must only be "patient" and wait for nucleosome breathing.

## References

1. S.B. Smith, L. Finzi, C. Bustamante, *Science*, 258, 1122–1126 (1992).
2. P. Cluzel, A. Lebrun, C. Heller, R. Lavery, J.L. Viovy, D. Chatenay, F. Caron, *Science*, 271, 792–794 (1996).
3. Lebrun, R. Lavery, *Nucl. Acid Res.*, 24, 2260–2267 (1996).
4. H. Yin M.D. Wang, K. Svoboda, R. Landick, S.M. Block, J. Gelles, *Science*, 270, 1853–1857 (1995).
5. M.D. Wang et al., *Biophys. J.*, 72, 1335 (1997).
6. C. Hodges L. Bintu, L. Lubkowska, M. Kashlev, C. Bustamente, *Science*, 325, 626–628 (2009).

## 10.2 DRUG DELIVERY CARRIERS: LIPOSOMES AND POLYMERSOMES

### 10.2.1 General Requirements for Targeted Drug Delivery Carriers

Advertisements for beauty and anti-aging creams highlight the fact that they contain liposomes. These liposomes, which are small vesicles of a few tens to a few hundred nanometers, are made of one or more lipid bilayers (Figure 10.5, Sections 6.3 and 10.3) and may encapsulate active species in their aqueous core, which are then supposed to penetrate the epidermis and release the active agents. Obviously, it is necessary that liposomes are not destroyed before they can penetrate, despite the massages performed by the user. If the use of liposomes in cosmetics can often be primarily seen as a marketing argument, their utility for pharmaceutical applications is less questionable.

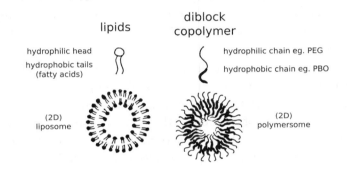

FIGURE 10.5  Liposomes and polymersomes.

The administration of drugs to target specific organs cannot always be done in a localized manner. On the other hand, direct administration into the body for targeted action assumes that during their journey towards the site of interest, the drugs are not degraded by specialized cells that detect foreign substances (such as macrophages) and that they will still be active upon arrival. To do this, a natural and simple strategy, in principle, consists in encapsulating drugs in nano-carriers, which are both (i) resistant or invisible to macrophages and (ii) degradable on-site to release their contents once the targeted organ has been reached. However, liposomes are relatively fragile; in particular, they are not very resistant to the shear imposed by blood flow. In this context, an alternative approach appeared in the 2000s and consisted of using polymersomes (Figure 10.5) instead of liposomes.

### 10.2.2 Stimuli-Responsive Polymersomes

The membrane of polymersomes is made of amphiphilic polymers that spontaneously associate in bilayers and close on themselves. The higher molecular weight of polymers compared to lipids leads to thicker membranes, inducing a greater resistance to flow, a greater impermeability, and a longer life time when they circulate in the blood (micro)vascularization [1]. However, this greater resistance should not become a disadvantage when the encapsulated active species must be released.

Since a passive release by diffusion through the membrane can be a very long process, it is important to easily and efficiently trigger the release by degradation, poration, or bursting of the polymersome. This is achieved by using the vast range of possibilities originating from the chemical synthesis of polymers.

Two main types of strategies were pursued. Variations in local physiological parameters, such as pH (which reaches a value of 2 in the stomach), temperature, oxidation (by reactive oxygenated species), specific enzymatic degradation, or application of an external stimulus such as light, magnetic field, ultrasound, or temperature [2, 3] can be used.

For example, in the first category, polymersomes made of diblock copolymers of poly(ethylene glycol)-b-poly (2,4,6-trimethoxybenzylidenepentaerythritol carbonate) show accelerated release at low pH.

Here, we will focus on the second category. A first example comes from S. Lecommandoux's group in Bordeaux. By co-encapsulating magnetic nanoparticles (of maghemite, $\gamma - Fe_2O_3$) with the therapeutic molecule of interest, they showed that an alternating magnetic field, oscillating at a frequency of 500 kHz, significantly increased the release kinetics, probably as a result of local overheating leading to an increase in permeability at the nanometric scale. Although some magnetic particles are biodegradable and non-toxic, a strong ethical concern often remains about the use of nanoparticles in the human body.

A second example, directly related to Section 5.2, comes from M.H. Li's group (in collaboration with two of the authors of this book). The idea is to use nematic liquid crystals that are sensitive to an external physical stimulus. In particular, as already seen in Section 9.2, azobenzene-based mesogenes undergo a *trans*-to-*cis* conformational change under irradiation using a 365 nm light. The M.H. Li group has synthesized a diblock copolymer, PEG-b-PMAazo444 ("PAzo"), which self-assembles into a polymersome (Figure 10.6). But, to our great surprise, prolonged illumination did not produce any significant effect other than slight vesicle crumpling. We had to adapt our approach and form asymmetric polymersomes, i.e. where the two sheets of the membrane are chemically different. The outer sheet was made of PAzo, sensitive to UV light, while the inner sheet was made inert, using PEG-b-PBD ("PBD") (Figure 10.6). Under UV illumination, all polymersomes, loaded with a dye, disappeared and released their colored content.

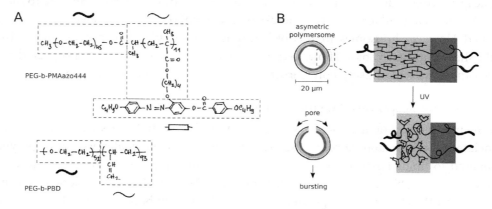

FIGURE 10.6 (A) Polymers used to produce light-responsive polymersomes (from M.H. Li group). PEG-b-PMAazo444 is UV-responsive. PEG-b-PBD is insensitive to UV illumination. (B) Opening of the polymersome under UV illumination and schematic zoom on the polymer conformations in the membrane.

To better visualize the bursting process, instead of preparing SUVs (~100 nm), we used giant vesicles, using a well-established reverse emulsion method [4]. These giant polymersomes can reach a diameter of 5 to 50 μm and are therefore clearly visible under an optical microscope. We observed that all our asymmetric vesicles burst and disintegrate in less than 300 ms when illuminated with a microscope lamp and a suitable 365 nm filter (Figure 10.7A). Although we cannot have direct access to the molecular scale within the membrane, we have been able to decipher the mechanism. Indeed, this experiment finds a direct analogy with a very simple experiment that illustrates the so-called "instability of curly hair." When a strip of tracing paper is placed on the surface of the water, it curves quickly at its ends and forms two cylinders of rolled paper (Figure 10.7B). This instability results from a bimorph effect caused by the swelling of the paper surface in contact with water, while the upper surface of the paper remains dry during the first few seconds because the coating of the tracing paper reduces the rate of water soaking. In the case of the polymersome membrane, the inner sheet, first in *trans* conformation and therefore stretched, undergoes an isotropic nematic transition, which leads (Figure 10.6B) to a decrease in the thickness of the liquid crystal polymer block and concomitantly, to an increase in the projected area of the outer sheet (due to volume conservation, a decrease in one dimension generates an increase in the other dimensions).

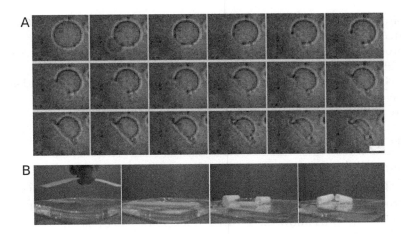

FIGURE 10.7 (A) Sequence of videomicrographs of a bright polymersome under UV illumination. Scale bar = 5 μm. In the second photograph, the expulsion of the internal liquid marks the initial moment of pore formation. One can note the formation of a "roll" of membrane towards the outside. (B) Macroscopic classroom experiment showing the curling instability of tracing paper at the water surface. (Extracted from [2].)

The membrane of the polymersome therefore consists of two sheets coupled to each other but with a significantly different area. This generates frustration within the membrane. Energetically, it becomes more favorable to bend the membrane to stretch the outer sheet (in excess of area) and compress the inner sheet (in lack of area compared to the median value). There is therefore a spontaneous curvature of the membrane, which has been modified by the change in conformation of the liquid crystal polymer.

Moreover, we can see in snapshots taken with a fast camera that once a pore has been nucleated (by accumulation of elastic energy), it opens by rolling *outwards* to relax the stored curvature energy.

As a validation, we obviously reversed the configuration. By placing the liquid crystal block in the inner sheet, this outward curling disappears. Although it is more difficult to see, we can guess that the curling is then going inwards.

In brief, this work shows how the properties of nematic liquid crystal polymers can once again be exploited to trigger the bursting of polymersome at a distance. This has to be considered as a proof of concept, since the dispersion of azobenzene groups would be toxic to the body. Chemical synthesis efforts to achieve biodegradable and non-toxic formulations are still needed. In addition, the use of light as an external stimulus limits applications to surface treatments, as biological tissues absorb a lot of light, especially in UV. While there are now more and more development strategies available to trigger the release of polymersome content, it should not be forgotten that it is also important to ensure that polymersomes will reach their targets. To do this, functionalization of the surface of polymersomes by ligands that specifically interact with receptors present on the surface of the target cells, i.e. to make them more "biomimetic," is generally the prevalent approach [3].

References
1. D.E. Discher, V. Ortiz, G. Srinivas, M.L. Klein, Y. Kim, D. Christian, S. Cai, P. Photos, F. Ahmed, *Prog. Polym. Sci.*, 32, 838–857 (2007).
2. E. Mabrouk, D. Cuvelier, F. Brochard-Wyart, M.-H. Li, P. Nassoy, *Proc. Nat. Acad. Sci. U.S.A.*, 7294–7298 (2009).
3. H. De Oliveira, J. Thevenot, S. Lecommandoux, *Wiley Interdiscip. Rev.: Nanomed. Nanobiotechnol.*, 4(5), 525–546 (2012).
4. S. Pautot, B.J. Frisken, D.A. Weitz, *Langmuir*, 19, 2870–2879 (2003).

## 10.3 BIOLOGICAL OR BIOMIMETIC MEMBRANES

Cells and intracellular compartments are delimited by membranes that combine the following three properties: (i) they serve as a protective barrier for the content of the cell or of the compartment, (ii) they allow the entry of nutrients and the disposal of waste, and (iii) they are flexible and thus allow the growth of a cell and its division. What are biological membranes made of? What are their most remarkable physical properties?

### 10.3.1 Composition of a Biological Membrane

Figure 10.8 shows a scheme of the plasma membrane of an animal cell. It is a mixture of about 50% lipids, 40% proteins, and 10% carbohydrate by mass. Carbohydrates are found in the glycocalyx which prevents undesired adhesion. Proteins come from two main families: those that are part of the cytoskeleton filaments underlying the membrane and anchored to it and those that constitute the membrane proteins themselves. Membrane proteins are often transmembrane proteins and can either diffuse freely in the membrane or be bound to the cytoskeleton via protein complexes. Ion channels and pumps that selectively allow ions to enter or exit the cell are among the most common membrane proteins. Adhesion proteins, which allow specific recognition of cells, are also part of this category.

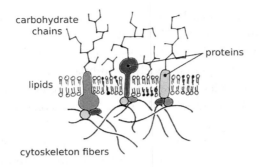

FIGURE 10.8    Drawing of a plasma membrane showing the main molecular components.

Since lipids are much smaller than proteins and carbohydrates ($M_{lipid} \sim 300$ g/mol $\ll M_{protein}$, $M_{carbohydrate} \sim 10^4–10^5$ g/mol), lipids are more numerous (about 50 lipids to one protein). Between 500 and 1,000 species of lipids have been recorded [1]. But they can be grouped into three main families (Figure 10.9) in decreasing order of concentration: (i) phospholipids, which comprise a glycerol molecule bound to two fatty acid chains and to a phosphate group carrying the polar residue; (ii) steroids (e.g. cholesterol), which contain a series of rigid aromatic rings and a small polar head (hydroxyl group -OH); and (iii) glycolipids, which comprise two fatty acid chains bound by an amino group and carry a sugar residue.

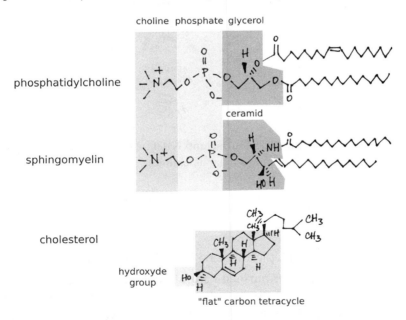

FIGURE 10.9    Chemical structure of the main three types of phospholipids, glycolipids, and steroids present in a plasma membrane.

As revealed by their chemical structure, lipids are amphiphilic molecules. The formation of supramolecular assemblies results from the competition between, on the one hand, steric and electrostatic repulsion between polar heads and, on the other hand, Van der

Waals attraction between hydrocarbon chains. However, unlike single-chain hydrocarbon surfactants, which mainly form micelles because of the conical shape of their envelope (Section 6.1), phospholipids, which are the most abundant lipids in biological membranes, have a cylindrical shape and spontaneously form bilayers, as set by a packing parameter $v/a_0l_c$ close to 1 (with $v$ the molecular volume, $a_0$ the projected area and lc the length of the molecule(Figure 10.10). Moreover, while the CMC (Section 6.1) of conventional surfactants such as sodium dodecyl sulfate (SDS) is in the range of $10^{-2}$ M, the CMC of phospholipids is around $10^{-9}$ M. As a consequence, phospholipids are virtually insoluble in water as monomers (individual molecules). They are all found as organized in bilayers. The presence of many different types of lipids in cell membranes may a priori lead to two-dimensional phase separations (Section 6.2). This has been the subject of intensive recent research, because lipid segregation at the nanoscale (i.e. not visible under an optical microscope) is the basic hypothesis for the concept of lipid rafts and proposes that areas rich in different types of lipids (and proteins) have specific cellular functions.

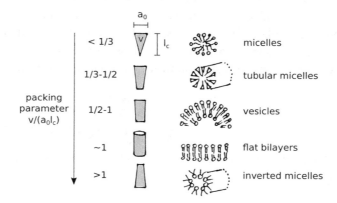

FIGURE 10.10    The packing parameter of surfactants is defined by the shape of the molecule ($v/a_0l_c$) and determines the structure of supramolecular assemblies.

To study the physical properties of biological membranes, a model system, the liposome, was widely used.

## 10.3.2 Physical Properties of Lipid Bilayers: Modeling and Experiments

Chemically synthesized lipids or purified natural lipids (such as egg yolk lecithin) are to make liposomes. By spreading them on a glass slide, followed by hydration, bilayers form spontaneously and close on themselves to form liposomes or vesicles. The slower the hydration process, the larger the vesicles can be, up to sizes of several tens of microns. They are then called giant vesicles (or "GUV" for giant unilamellar vesicles). They are of practical use because they can be seen with an optical microscope. Another method for obtaining GUVs is the so-called electroformation method. Lipids are spread on a conductive plate, and an alternating electric field (at about 10 Hz) is applied. This promotes lipid detachment and accelerates the formation of GUVs.

### 10.3.2.1 Fluidity

Lipid bilayers are generally fluid at room or body temperature (37°C). This property is important for cellular functions. Lipids and proteins can diffuse within the plane of the membrane. Their diffusion coefficients are measured by fluorescence recovery after photobleaching (FRAP) (Figure 10.11). This method consists of using a bilayer doped with fluorescent lipids. By irradiating an area of characteristic membrane size $\ell$ with a high intensity laser beam, fluorescence is extinguished. At time $t=0$, there is a black area in a bright background. If lipids diffuse, fluorescence will be re-homogenized. The measurement of the characteristic time $\tau$ of fluorescence recovery allows us to derive the diffusion coefficient: $D_{\mathrm{lipid}} \sim \ell^2 / \tau \sim 1~\mu m^2/s$.

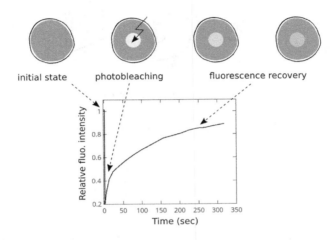

FIGURE 10.11    Principle of the FRAP method to measure the lipid diffusion coefficient.

Lipids can also transit from one leaflet of the bilayer to another. This so-called flip-flop process is for most lipids (one occurrence every $10^8$ s) in vesicles. In cells, specialized proteins called flippases (flipping a lipid from the external leaflet to the internal one) or floppases (flipping a lipid in the opposite direction) can speed it up and create an asymmetry of lipid composition between the two leaflets of the membrane, and scramblases can dissipate this asymmetry.

When the temperature of a lipid bilayer is lowered, its fluidity decreases. We move from a two-dimensional liquid state to a liquid crystal state (Sections 5.1 and 5.2).

### 10.3.2.2 Mechanical Properties

The first theoretical works on the mechanical properties of membranes date back to the 1970s [2, 3, 4]. The three types of deformation of a membrane or a sheet are extension (or conversely compression), shear, and bending (Figure 10.12). Each of these deformations is associated with an energy. Membranes, which generate an interface between two aqueous media, are also characterized by a surface tension, called membrane tension. All these physical parameters will be detailed hereafter.

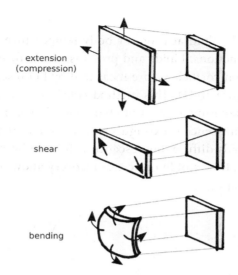

FIGURE 10.12    Primary modes of deformation of a membrane.

*Extension/Compression:* The energy per unit area associated with the extension is given by:

$$H_{ext} = \frac{1}{2}\chi\left(\frac{\Delta A}{A}\right)^2 \tag{10.1}$$

We recognize the general expression of the elastic energy stored by a spring (Hooke's law) and extended here to two dimensions. The compressibility modulus $\chi$ is the 2D analog of the stiffness of a linear spring, and the relative surface variation $\Delta A/A$ is the analog of the relative elongation. $\Delta A / A$ does not exceed ~5%. Beyond this value, there is lysis of the membrane. The measured compressibility moduli (see Figure 10.13) are ~0.1 J/m² or N/m.

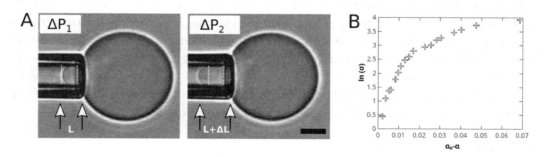

FIGURE 10.13    (A) Microphotography of a vesicle aspirated into a micropipette. $L$ is the length of the aspirated tongue. (B) Plot of surface tension as a function of the excess surface area of a vesicle aspirated in a pipette on a semi-logarithmic scale (to emphasize the entropy regime at low tensions).

*Shear:* A fluid membrane does not offer any resistance to shear. This term is therefore generally neglected. It only becomes relevant for red blood cells, for example, for which the membrane is supported by a spectrin network. When red blood cells sneak into narrow blood capillaries (Section 2.2), the required shear deformation is mainly assigned to this spectrin network.

*Bending:* The bending energy density is given by:

$$H_{bend} = \frac{1}{2}\kappa\left(c_1 + c_2 - c_0\right)^2 + \kappa_G c_1 c_2 \tag{10.2}$$

where $c_1 = 1/R_1$ and $c_2 = 1/R_2$ are the two principal curvatures of the membrane (see Section 4.2), $c_0$ is the spontaneous curvature, $c_1 c_2$ is the Gaussian curvature, $\kappa$ the bending modulus (with the dimensions of an energy), and $\kappa_G$ the Gaussian bending modulus. The situation is often simplified, because, according to Gauss–Bonnet's theorem, the Gaussian curvature integrated over the surface of the vesicle is a topological invariant. So, if the membrane is deformed without making holes in it for example, the Gaussian curvature does not change, and there is no need to consider it explicitly. The spontaneous curvature represents a possible asymmetry in the composition of the two leaflets of the membrane. An example where spontaneous curvature is important and used in practice is illustrated in Section 9.2. But, for biomimetic GUVs, $c_0 \approx 0$. The bending moduli can be measured (see Figure 10.13) and are in the order of $10\,k_BT$ ($\sim10^{-20}$ J).

*Membrane Tension:* When a membrane of surface area $A$ is stretched, its area varies by $\Delta A$. The associated surface energy is therefore: $H_{tens} = \sigma\dfrac{\Delta A}{A}$ where $\sigma$ is the membrane tension. According to Equation 1, we have:

$$\sigma_H = \chi\frac{\Delta A}{A} \tag{10.3}$$

Here, the origin of the tension is enthalpic (hence the $H$ index). This relation is valid if the vesicle is already tense.

Often, vesicles and cells are in an unstressed and floppy state. They deform under the effect of thermal agitation; they undulate and cause the scattered light to flicker (this flicker effect remained "magical" for a long time and was eventually explained for red blood cells in terms of membrane fluctuations by F. Brochard and J.-F. Lennon in 1975 [5]). So, when pulling on a fluctuating vesicle, we start by smoothening out the fluctuations of the membrane, which can be seen as nanometric wrinkles leading to a decrease of the excess area $A - A_p$, where $A$ is the area of the undulated membrane and $A_p$ is the projected area. Helfrich has shown that the entropic tension $\sigma_E$ increases exponentially as the entropic ripples are unfolded:

$$\sigma_E \approx \frac{\kappa}{\ell^2}\exp\left(-\frac{8\pi\kappa}{k_BT}\frac{A - A_p}{A}\right), \tag{10.4}$$

where $\ell$ is a microscopic cut-off length.

In the general case, Evans [6] took into account both the enthalpic and entropic contributions and related the excess area $\Delta\alpha = \left(\dfrac{\Delta A}{A}\right)_{\sigma=0} - \left(\dfrac{\Delta A}{A}\right)_{\sigma}$ to the two main moduli $\chi$ and $\kappa$:

$$\Delta\alpha = \frac{k_B T}{8\pi\kappa}\ln(1+cA\sigma)+\frac{\sigma}{\chi}, \tag{10.5}$$

with $c = 1/24\pi$.

### 10.3.2.3 Vesicle Micropipette Aspiration Experiments

The mechanical parameters defined in the previous paragraph can be measured by forcing the vesicles to change morphology under the application of controlled mechanical constraints. The main method, developed by E. Evans [6], consists of aspirating a giant vesicle into a micropipette. To do this, a pressure difference ($\Delta P > 0$) is imposed between the liquid medium and the inside of the pipette. Upon each pressure increment, the vesicle penetrates a little more into the pipette and increases the aspirated length $L$ (or "tongue") (Figure 10.13). The "sphere + tongue" geometry allows the simple measurement of the excess area $\Delta a$, and the pressure $\Delta P$ sets the membrane tension $\sigma$. Indeed, the application of Laplace's law (Section 4.2) between the inside of the vesicle tongue and the pipette on the one hand and between the spherical portion of the vesicle and the external solution on the other hand simply gives:

$$\sigma = \frac{\Delta P}{2\left(\dfrac{1}{R_p}-\dfrac{1}{R_v}\right)}, \tag{10.6}$$

where $R_p$ is the radius of the pipette, and $R_v$ is the radius of the vesicle.

The very narrow micropipette allows the amplification of the variations of surface area. The size of the spherical portion of the vesicle is almost constant. So, the decrease of the excess area is given by: $\Delta\alpha = 2\pi R_p \Delta L$, where $\Delta L$ is the increase of the length of the tongue.

### 10.3.2.4 Permeability

The permeability of a membrane for a molecule can be quantitatively assessed by its permeability coefficient $P$ (in cm/s), which is a measure of the speed at which the molecule can cross the membrane. The flow of molecules (in number per second and per unit area) depends on the difference in concentration between the two sides of the membrane. $P$ is given by the ratio of this flow to the difference in concentration. In general, a membrane composed solely of lipids is almost impermeable to sodium, potassium, and chlorine ions (Figure 10.14). In a biological context, intra- and extra-cellular concentrations are finely regulated. Ion entry and exit can only be achieved through transport processes, via pumps and ion channels. On the other hand, water permeability is the highest. It is also important that in the event of osmotic shock, the cell can restore the osmotic balance it needs to survive.

Note that these permeability coefficients can also be derived from micropipette experiments by measuring the swelling (or shrinking) of the vesicle when it is suddenly immersed in a hypo-osmotic (or hyper-osmotic) medium [7].

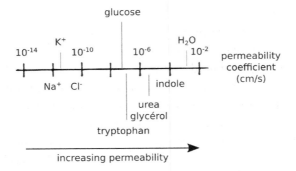

FIGURE 10.14   Passive permeability coefficients through a lipid bilayer for a few molecular species.

### 10.3.3 Morphologies of Lipid Vesicles

#### 10.3.3.1 Phase Diagram

It is striking to note that there is large morphological variety in a population of vesicles. The important parameter that determines the shape of a vesicle is the surface-to-volume ratio. The volume $V$ can easily be controlled by tuning the osmolarity. If the total volume is close to $4\pi R^3/3$ (maximum volume for the given area $A$), i.e. for a reduced (dimensionless) volume $v$ close to 1, the vesicle will be almost spherical. If $v$ decreases, by deflating the vesicle, ellipsoidal, biconcave disc, or pear-like shapes may appear. From a theoretical point of view, if the difference in surface area between the two leaflets, noted $\Delta a_0$ (related to spontaneous curvature), is further varied, a complete phase diagram can be drawn (Figure 10.15), where most morphologies have been experimentally reported.

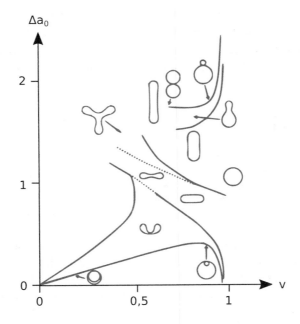

FIGURE 10.15   Phase diagram of vesicles in the space of parameters defined by the reduced volume and the difference in effective surface area between leaflets (adapted from [8]).

### 10.3.3.2 Multi-Component Vesicles

When several types of lipids are mixed to form vesicles, phase separations may occur (Figure 10.16). A domain composed mainly of one type of lipid floating in an "ocean" consisting mainly of another type of lipid can be seen as the analog of a drop of an immiscible solvent in another, with the difference that it is here a "two-dimensional drop." The equivalent of the surface tension in 3D becomes the line tension in 2D. This reflects the energy cost associated with the creation of a domain in the membrane. If the vesicle is very tense or has a high modulus of curvature rigidity, segregated areas will be seen in the vesicle plane. However, if the tension is low and/or if the bending modulus is low, the domains can bud and give rise to various shapes (Figure 10.16) [9].

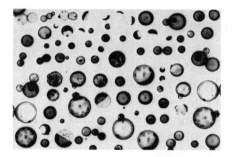

FIGURE 10.16   Fluorescence microscopy images of vesicles with coexistence between two phases. (Courtesy of Aurélien Roux.)

### 10.3.3.3 Nanotubular Shapes

The action of external mechanical stress on a vesicle can also trigger a shape change. If we apply a point force, a membrane thread, or more precisely, a tube with a sub-micrometer diameter, appears. For example, this force can be applied by bringing an adhesive bead in contact with the vesicle (held in a pipette as in Figure 10.17) and attempting to separate it with an optical trap.

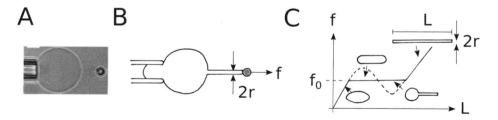

FIGURE 10.17   (A) Phase contrast microscopy image showing the aspirated vesicle, the bead manipulated with an optical trap, and the nanotube in between. (Extracted from [10].) (B) Drawing of the experiment leading to the formation of a membrane nanotube. (C) Schematic graph of the force $f$ as a function of the elongation $L$ of a vesicle of initial radius $R_0$ and tension $\sigma$. The dotted line indicates an unstable region and $f_0$ is the tube extrusion force. (Adapted from [11].)

The membrane is fluid. However, from everyday life observations, we know that a cylinder made of water is not stable. Indeed, everyone has already seen that the water jet that flows from our tap breaks down into small droplets. This is due to the Plateau–Rayleigh instability: a liquid cylinder has a surface, therefore an interfacial energy (with air), that is higher than a large number of small droplets with the equivalent total volume. This instability also explains why the dew on spider webs does not form a continuous sheath but small droplets attached to the web. However, in the case of fluid membranes, a cylinder is stable. Qualitatively, the destabilizing effect of membrane tension (which would tend to make small spherical vesicles) is balanced by the stabilizing effect of bending energy (which tends to avoid the formation of these small vesicles, which are "more curved" than a cylinder). As explained in Figure 10.17C, the deformation of a spherical vesicle into an ellipsoid becomes energetically too costly compared to a situation where a sphere coexists with a tube. Remarkably, before the entire membrane is transformed into a tube, the elongation is performed at constant force. We can show that the threshold $f_0$ required to pull a membrane tube is $f_0 = 2\pi\sqrt{2\kappa\sigma}$. For typical values of $\kappa \sim 10\ k_B T$ and $\sigma \sim 10^{-4}$–$10^{-6}$ N/m, we find that $f_0$ is about a few tens of picoNewtons, which is in agreement with measured values. In parallel, the radius of the tube is given by $r = \sqrt{(\kappa/2\sigma)}$, which gives values for $r$ in the 10 to 100 nm range.

These nanotubes are a simple way to measure the bending rigidity of a membrane. Such membrane threads have also been observed in animal cells, suggesting that this exotic membrane shape may have a role in remote intercellular communications.

### References

1. M. Edidin, *Nat. Rev. Mol. Cell. Biol.*, 4, 414–418 (2003).
2. P.B. Canham, *J. Theor. Biol.*, 26, 61–81 (1970).
3. W. Helfrich, *Z. Naturforsch. C*, 28, 693–703 (1973).
4. E.A. Evans, *Biophys. J.*, 13, 926–940 (1973).
5. F. Brochard, J.-F. Lennon, *J. Phys. France*, 36, 1035–1047 (1975).
6. E. Evans, W. Rawicz, *Phys. Rev. Lett.*, 23, 2094–2097 (1990).
7. W. Rawicz, K.C. Olbrich, T. McIntosh, D. Needham, E. Evans, *Biophys. J.*, 79, 328–339 (2000).
8. Y. Sakuma, M. Imai, *Life*, 5, 651–675 (2015).
9. T. Baumgart, S. Hess, W.W. Webb, *Nature*, 425, 821–824 (2003).
10. D. Cuvelier N. Chiaruttini, P. Bassereau, P. Nassoy, *Europhys. Lett.*, 71, 1015–1021 (2005).
11. O. Rossier D. Cuvelier, N. Borghi, P.H. Puech, I. Derényi, A. Buguin, P. Nassoy, F. Brochard-Wyart, *Langmuir*, 19, 575–584 (2003).

## 10.4 CYTOSKELETAL POLYMERS

The cytoskeleton is a network of protein filaments that contribute to cell integrity and structural rigidity, regulate cell morphology and shape changes, and exert forces to contribute to cell motility.

## 10.4.1 Different Types of Filaments

Cytoplasm refers to the intracellular fluid in which the various organelles (mitochondria, Golgi apparatus, lysosomes) and polymers of the cytoskeleton are dispersed. These biopolymers are grouped into three families: actin filaments, microtubules, and intermediate filaments (Figure 10.18). They differ in particular by:

- The chemical composition (i.e. the protein nature of their monomers).

- The width (or apparent diameter) of the filaments: from 7 nm for actin to 25 nm for microtubules.

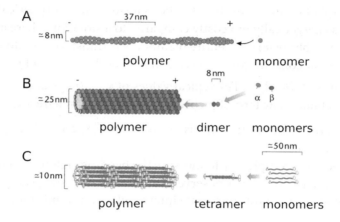

FIGURE 10.18   Drawing of (A) actin filaments, (B) microtubules, and (C) intermediate filaments and their respective subunits ("monomers").

### 10.4.1.1 Actin (Figure 10.19A)

Actin filaments (or F-actin) filaments are composed of globular actin (G-actin) monomers that can be combined with a molecule of ATP (adenosine triphosphate) or ADP (adenosine diphosphate), derived from ATP by hydrolysis. They are all associated in the same direction. This gives a polarity to the filament which has the shape of a double helix. We define a "barbed" or "plus" end, rich in monomers bound to ATP, and a "pointed" or "minus" end, rich in monomers bound to ADP.

### 10.4.1.2 Microtubules (Figure 10.19B)

Microtubules are built from tubulin units. The monomers, which are dimers of α- and β-tubulin, are organized into (linear) protofilaments. Thirteen of these protofilaments then combine into hollow cylinders. Tubulin monomers have binding sites for guanosine triphosphate (GTP) or its hydrolyzed form, guanosine diphosphate (GDP).

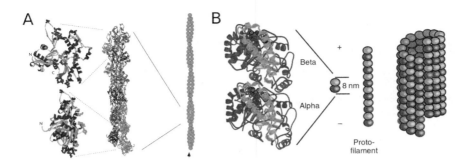

**FIGURE 10.19** Monomers (A) of actin filaments and (B) of microtubules. The crystallographic representation shows α helixes and β sheets, which are the two forms of secondary protein structure.

### 10.4.1.3 Intermediate Filaments

Intermediate filaments contain several classes of filaments such as vimentin, desmin, keratin, and lamin. They are dimers of polypeptides that combine into tetramers in an antiparallel manner, then into protofilaments by end-to-end association. Finally, about eight protofilaments form an intermediate filament that has a "rope" structure. They are not polar and much less dynamic than microtubules and actin filaments. They essentially contribute to the morphological, structural, and elastic properties of the cell. They have been less studied than actin filaments and microtubules.

## 10.4.2 Rigidity

Unlike many synthetic polymers or DNA in solution, which adopt a coil configuration, actin filaments, microtubules, and intermediate filaments look like (slightly undulating) rods. They are semi-flexible polymers. In general, the rigidity of a polymer is characterized by its persistence length $L_p$ (Section 7.2) in thermodynamic equilibrium.

### 10.4.2.1 Relationship between Persistence Length and Rigidity

Let us consider a thin flexible rod of fixed length $L$ and radius $b$, subjected to thermal forces. Its shape is completely determined by the tangent vector $\vec{t}(s) = d\vec{r}(s)/ds$ or equivalently by the tangent angle $\theta(s)$ along the rod, with $s$ the curvilinear abscissa (Figure 10.20). The tangent correlation function is defined by: $g(s) = \langle \vec{t}(s) \cdot \vec{t}(0) \rangle$. For $s \approx 0$, $g(s) \to 1$ (correlation) and for $s \gg L_p$, $g(s) \to 0$ (decorrelation). Such properties are satisfied by an exponential function. $L_p$ is thus defined as the characteristic arc length beyond which thermal fluctuations of the angle $\theta(s)$ become uncorrelated.

$L_p$ is also directly related to the rod flexural rigidity $\kappa_f$ defined by $\kappa_f = E \cdot I$, where $E$ is the material Young's modulus (in Pascal) and $I = \iint_{\text{section}} y^2 dA$ is the geometric moment of inertia that characterizes the shape of the rod. The bending energy of a rod is given by:

$$E_c = \frac{\kappa_f}{2} \int_0^L \frac{1}{r(s)^2} ds$$

with $r(s) = ds/d\theta$.

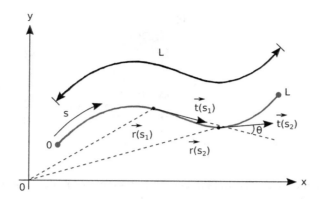

FIGURE 10.20  Thermal fluctuations of a thin rod.

We thus obtain:

$$E_c = \frac{\kappa_f}{2} \frac{\langle \theta \rangle^2}{L}$$

where $\theta$ is the angle averaged over the length of the rod.

Under thermal energy, we have: $E_c \sim k_B T/2$, which leads to $\langle \theta \rangle^2 = (k_B T/\kappa_f)L = L/L_p$. This defines the persistence length as:

$$L_p = \frac{\kappa_f}{k_B T} = \frac{EI}{k_B T} \sim \frac{E}{k_B T} b^4 \sim \frac{b^4}{a^3}$$

with $a$ the monomer size. This strong dependence of $L_p$ on $b$ partly explains why a moderate increase of the apparent (projected) radius of DNA when going from single strand to double strand leads to an increase from 2 nm to 50 nm of the persistence length. We might also keep in mind that hairs of radius 10 µm and Young's modulus 7 GPa have a persistence length of about 10,000,000 km. They only bend because of gravity. In a zero-gravity environment, they would be perfectly straight.

### 10.4.2.2 Experimental Measurements of the Persistence Length

Experimentally, the persistence lengths of actin filaments and microtubules were measured by analyzing their thermal fluctuations, recording filament shapes, and averaging over all configurations (Figure 10.21A). Another method, which was developed for microtubules, consisted of attaching the microtubule at one end to an axonema (used as a handle to secure the growth of microtubules that can form cilia or flagella) and manipulating the other end with an optical trap (Figure 10.21B). After deflecting the microtubule, the laser is switched off, and the microtubule end relaxes towards its equilibrium position. This movement is described by the balance between the elastic force of a beam, $f_e = \kappa_f (d^4 y/dx^4)$, and the viscous force applied to a moving cylinder of length $L$ and radius $b$, $f_v = [2\pi/\ln(L/b)] \cdot \eta V L$, where $V$ is the velocity of the cylinder at its free end, and the first term is a numerical coefficient related to the cylindrical geometry. The relaxation time constant $\tau$, which can be exactly calculated, is proportional to the ratio of the elastic constant and the hydrodynamic friction coefficient, i.e. $\tau \cong \kappa_f/\eta \propto L_p$.

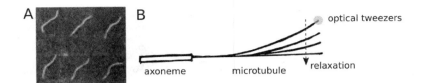

FIGURE 10.21 (A) Images of actin filaments by fluorescence microscopy (taken at 10 s intervals) and showing thermal fluctuations. (Extracted from [1].) (B) Experimental setup designed to measure the bending rigidity of a microtubule using an optical clamp (asterisk) that allows the end of the microtubule to be deflected from its resting position. (Adapted from [2].)

All these experiments show that $L_p \approx 10$–$20$ μm for actin filaments and $L_p \approx 1$–$6$ mm for microtubules.

There are two orders of magnitude of difference between the persistence lengths of actin and microtubules. Where does this difference come from? The geometric moment of inertia of a microtubule can be calculated by considering it as a hollow cylinder of internal radius $r_i = 9.5$ nm and external radius $r_e = 12.5$ nm, while actin can be considered as a plain cylinder with $r_i = 0$ and $r_e = 3.5$ nm:

$$I = \iint_{section} y^2 dA = \int_{r_i}^{r_e} \int_0^{2\pi} (r\sin\theta)^2 dr \cdot r d\theta = \frac{\pi}{4}\left(r_e^4 - r_i^4\right)$$

$I_{actin} = 1.2 \times 10^2$ nm$^4$ and $I_{microtubule} = 1.28 \times 10^4$ nm$^4$.

Since we find the same 100-fold difference, this means that the Young's modulus $E$ of the tubulin "material" and of the G-actin "material" are similar, which is not so surprising since both are proteins, i.e. peptide assemblies. The stiffness of the polymers of the cytoskeleton therefore originates from their shape. In addition, an estimate of the Young's modulus of the protein "material" gives $E \sim 1$–$4$ GPa, which, counter-intuitively, is higher than that of a polyethylene block.

## 10.4.3 Dynamics

Actin filaments and microtubules have the peculiarity of being very dynamic (unlike intermediate filaments). This property is crucial with regard to their cellular functions. It is described in detail in [3].

### 10.4.3.1 Actin Treadmilling

Let us start with actin, which is the most abundant protein in most cells (at concentrations of several grams per liter). A cell, which is spread on a substrate, looks like a fried egg and has a vaguely circular outline. However, in the presence of a chemo-attractant gradient, it polarizes and begins to move towards the source. This polarization and the subsequent cellular motility are partly induced by the polymerization of actin filaments. If the source of the chemoattractant changes position, these filaments quickly disassemble and reform in the new direction (Figure 10.22). To understand the phenomena that regulate actin assembly and disassembly dynamics, it is necessary to study the polymerization kinetics of these biopolymers, which is inherently reversible because it is based on weak (non-covalent) interactions.

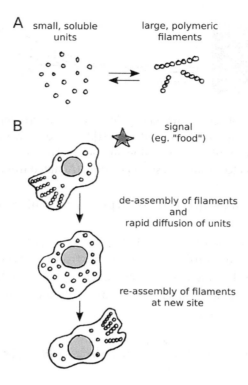

FIGURE 10.22   Importance of the polymerization and depolymerization of actin filaments in cell motility.

In general, polymerization and depolymerization can be considered as first-order chemical reactions where a monomer ($P_1$) is added to a polymer chain of $n$ or $n-1$ units (denoted $P_n$ or $P_n-1$, respectively) to elongate it by one unit. Similarly, when a monomer comes off the end of the polymer chain, it is shrunk by one unit. This can be written as:

$$P_n + P_1 \rightleftarrows P_{n+1} \tag{a}$$

$$P_{n-1} + P_1 \rightleftarrows P_n \ (n>1) \tag{b}$$

with $k_{on}$ and $k_{off}$ the association and dissociation kinetics constant such that:

$$\frac{d[P_n]}{dt} = k_{on}[P_{n-1}]\cdot[P_1] + k_{off}[P_{n+1}] - \left(k_{on}[P_n]\cdot[P_1] + k_{off}[P_n]\right) \tag{10.1}$$

This relationship expresses the rate of production of $P_n$ ([] indicates the concentration of the species) by writing the balance between production and consumption of the species.

The average number of monomers in the polymer chain is obtained by taking the average for all n values of the probabilities to obtain an n-mer, $P_n$. This probability is directly proportional to the concentration:

$$\langle n \rangle \approx \sum_{n=1}^{\infty} n[P_n] \tag{10.2}$$

By taking the derivative with respect to time and using Equation 1, we obtain, after simplification:

$$\frac{d\langle n \rangle}{dt} = k_{on}[P_1] - k_{off} \tag{10.3}$$

Thus, by defining $K_d = k_{off}/k_{on}$, the equilibrium dissociation constant, we find:

- For $[P_1] < K_d$, the average length of the filaments decreases.
- For $[P_1] > K_d$, the average length of the filaments increases.

The local concentration of G-actin therefore controls the polymerization or disassembly of actin filaments. $K_d = c^*$ is the critical concentration.

But in the case of a polar actin filament, we saw that the "+" end was rich in ATP-bound monomers and the "−" end was rich in ADP-bound monomers. Therefore, two different polymerization/depolymerization reactions should be considered at both ends (Figure 10.23A).

$$\frac{d\langle n^+ \rangle}{dt} = k_{on}^+[P_1] - k_{off}^+ \quad \text{and} \quad \frac{d\langle n^- \rangle}{dt} = k_{on}^-[P_1] - k_{off}^-$$

This defines two critical concentrations $c_+^* = (k_{off}^+/k_{on}^+)$ and $c_-^* = (k_{off}^-/k_{on}^-)$. As shown in the graph in Figure 10.23B, there is a concentration $c_{TM}$ for which the polymerization rate of the + end is equal to the polymerization rate of the − end, which means that the ends are constantly renewed, but the filament length remains constant. If a monomer initially located in the middle of the chain is tracked, it has an apparent movement towards the − end, hence the name "treadmilling."

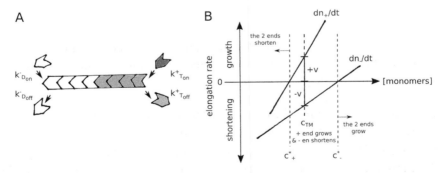

FIGURE 10.23 (A) Drawing of polymerization and depolymerization reactions at both ends of an F-actin chain. (B) Graph of the elongation rate as a function of monomer concentration, highlighting the existence of a "Treadmilling" concentration.

### 10.4.3.2 Dynamic Instability of Microtubules

Like actin filaments, microtubules are characterized by a peculiar assembly in which GTP-tubulin dimers are mainly found at the + end while dimers bound to GDP are located at the − end (Figure 10.24A). But the dynamics are completely different. Filament growth is achieved by adding GTP dimers to the + end. Over time, GTP is hydrolyzed to GDP. The monomer units located far from the + end are the "oldest" ones, so they have a high probability of being linked to GDPs. On the other hand, the cap of the + end is essentially composed of GTP-bound monomers. In the presence of a dimeric tubulin reservoir at concentration $c_0$, we observe dynamics marked by cycles of constant growth phases followed by catastrophes corresponding to a drastic collapse by depolymerization (Figure 10.24B). These dynamics are explained by the fact that the GTP cap is essential to stabilize the end of the microtubule. So, if the hydrolysis rate of the GTPs, $v_h = a / \tau$ (with $a$ the size of a dimer and $\tau$ the hydrolysis time) is larger than the polymerization rate, the length of the GTP cap is reduced. According to Equation 3, we have directly:

$$v_p = a\frac{dn}{dt} = a \cdot k_{\text{on}} \left( c_0 - n(t) \cdot \frac{G}{V} \right),$$

where n is the number of dimeric units in an average filament, G is the number of pre-formed filaments (or germs) at time $t=0$, and V is the volume of the solution. Here, the dissociation of monomers via the kinetic constant $k_{off}$ is neglected.

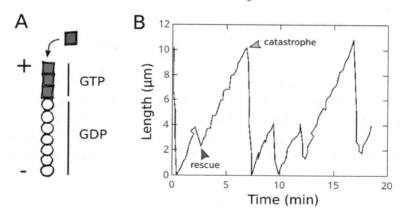

FIGURE 10.24 (A) Drawing of a microtubule filament indicating GTP monomers and GDP monomers. (B) Graph of the length of microtubules as a function of time showing the existence of catastrophic depolymerization. (Adapted from [4].)

By writing $v_h = v_p$ and solving the first-order differential equation, we obtain the expression of the critical time at which a catastrophe occurs:

$t_{\text{crit}} = (V/Gk_{\text{on}})\ln(\tau k_{\text{on}} c_0)$. The more concentrated the reservoir, the longer the time required for monomer depletion, so the longer the time between two catastrophes. Here, the model is deterministic and therefore sets a value for the catastrophe time. Experimentally, we observe a distribution of critical times.

This dynamic instability is of major biological interest in the cell cycle. During cell division, the microtubules, which grow from the poles of the mitotic spindle (Figure 10.25), must capture the chromosomes by binding to the kinetochores, which are protein assemblies close to the center of the chromosomes. It is therefore a question of targeting a quasi-punctual area with a rod (one-dimensional) in the intracellular space. This is not easier than finding a needle in a haystack. For this capture to be effective, it is necessary to be able to make "trial and error" cycles. If the growing microtubule misses its target and continues to grow, it will have no chance of touching the kinetochore. On the other hand, it can be expected that a process that shortens it quickly to get another opportunity to hit the target will be optimal.

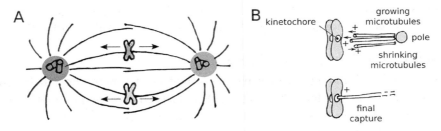

FIGURE 10.25   (A) During cell division, the chromosomes are captured at the level of the kineto-chore by microtubules before the mother cell is separated into two daughter cells. (B) Focus on the capture process.

References
1. A. Ott, M. Magnasco, A. Simon, A. Libchaber, *Phys. Rev. E*, 48, R1642–R1645 (1993).
2. H. Felgner, R. Frank, M. Schliwa, *J. Cell Sci.*, 109, 509–516 (1996).
3. R. Phillips, J. Kondev, J. Thériot, H.G. Garcia, *Physical Biology of the Cell*, 2nd Edition. Garland Science, 2013.
4. D.K. Fygenson E. Braun, A. Libchaber, *Phys. Rev. Lett. E*, 50, 1579 (1994).

## 10.5 BIOLOGICAL TISSUES AND ACTIVE SOFT MATTER

Studies in biology and biophysics on individual cells spread over the bottom of a petri dish have allowed the dissection of numerous mechanisms, which are crucial to the function and fate of a cell as an elementary entity. However, is it sufficient to understand the behavior and the evolution of biological tissues or organs, which are composed of a large number of cells? The importance of collective and/or cooperative effects is well known in physics. In biology, these are called emergent properties and reflect the fact that an assembly of cells is likely to have radically different properties from those of the "elementary brick," i.e. the single cell.

Many physics works inspired by the concepts and approaches used in soft matter have been conducted in the last decades to study collective properties of cells. Instead of working directly on tissues or organs, in vitro model systems were developed and extensively used. In particular, multicellular spheroids (Figure 10.26) can be formed by a technique derived from the flavor pearl method (Section 8.2) or by letting cells divide on a non-adherent substrate, which spontaneously drives aggregation. A multicellular spheroid typically contains a thousand cells for a size of a few hundred microns in diameter. Spheroids are now considered good models of three-dimensional cell cultures, making it possible to replace or reduce animal tests and avoid some artifacts associated with a two-dimensional culture in a petri dish.

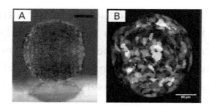

FIGURE 10.26 **(See color insert.)** A multicellular spheroid seen by (A) phase contrast microscopy and (B) confocal fluorescence microscopy. (Courtesy of S. Douezan (A); F. Bertillot (B).)

## 10.5.1 Multicellular Aggregates as Liquids

The formation of a cellular aggregate is based on the fact that interactions between neighboring cells (particularly induced by cadherins) are favored over interactions between cells and the substrate (e.g. between integrin and fibronectin). The shape that minimizes the energy of a liquid is the sphere. The fact that multicellular aggregates form spheroids is therefore a first indication that these microtissues have a liquid-like behavior.

### 10.5.1.1 Surface Tension of Tissues

If cellular spheroids have liquid properties, a surface tension is expected to be defined and measurable (Section 4.2). The classical experiment, first proposed by M. Steinberg [1], consists of a direct application of Laplace's formula (Section 4.2). By definition, in the general case of a liquid whose surface is characterized by two radii of curvature $R_1$ and $R_2$, the surface tension $\gamma$ is given by:

$$\gamma = \frac{\Delta P}{\left(\dfrac{1}{R_1} + \dfrac{1}{R_2}\right)}$$

where $\Delta P$ is the pressure inside the spheroid due to its curvature.

By squeezing the spheroid between two parallel plates (Figure 10.27) and measuring the force $F$ with a microbalance [2] or from the deflection of the most flexible of the two plates [3], and thus the pressure that flattens the spheroid, $\Delta P = F/\pi R_3^2$ where $R_3$ is the radius of the contact area between the plate and the spheroid, the surface tension is derived:

$$\gamma = F/\pi R_3^2 \cdot \left(\frac{1}{R_1} + \frac{1}{R_2}\right)^{-1}$$

The measured surface tensions of tissue aggregates are in the order of a few mN/m, i.e. an order of magnitude lower than the interfacial energy between water and air but at least a thousand times higher than the membrane tension of a liposome (Section 10.3).

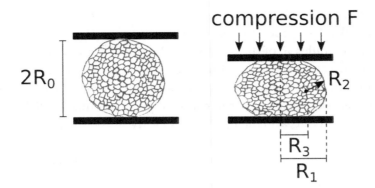

FIGURE 10.27  Cellular aggregates squeezed between two plates.

### 10.5.1.2 Differential Adhesion Hypothesis

The apparent surface tension of cellular aggregates is a quantitative signature of cell cohesion. Another consequence of these liquid properties for biological tissues has led to the differential adhesion hypothesis (DAH), proposed by P. Townes and J. Holtfreter more than 60 years ago [4]. DAH postulates that cells reorganize to maximize cohesive interaction energies within the aggregate and minimize their interfacial free energy.

This hypothesis was demonstrated by Steinberg in a simple and elegant experiment [2]. If two types of cells are mixed, those with a lower surface tension will gather (segregate) and coat those with a higher tension, in analogy with drops of immiscible liquids (Figure 10.28).

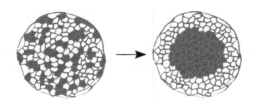

FIGURE 10.28  Sorting of "dark" and "bright" cells within a mixed aggregate.

DAH has recently been refined by introducing a contribution of cortical cell tension [5]. But the basic principle remains valid and observed experimentally.

### 10.5.1.3 An Active Shivering Liquid

Similarly to vesicles for which the membrane tension was measured by suction using micropipettes (see Section 10.3), cellular aggregates are aspirated into pipettes (Figure 10.29) [6]. When the suction pressure exceeds a threshold, the spheroid is gradually aspirated by forming a tongue inside the pipette. There is a rapid penetration regime followed by constant velocity penetration. More precisely, from the curve of the length of the tongue $L(t)$, we derive that the aggregate has a viscoelastic response. Fitting the experimental curve with the model allows the derivation of values for the surface tension, of the order of a few mN m$^{-1}$, of the elastic modulus $E$, of the order of 1 kPa, and of the viscosity $\eta \sim E\cdot\tau$ where $\tau$ is the characteristic time of the tissue ($\sim$ 1 h). There are two differences from the liquid drop model. First, $\gamma$ is not strictly constant but increases with $P$, which shows that the tissue is active and resists applied forces. Second, in a small pressure range $P \sim 0.5$ to 1 kPa, there is a shivering of the aggregate, also interpreted as an active response from the cell.

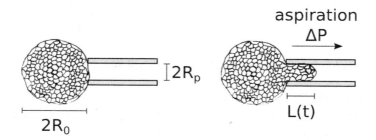

FIGURE 10.29  Micropipette aspiration of aggregates.

## 10.5.2 Wetting of Multicellular Aggregates

As seen previously, the cohesion of the aggregates is mainly ensured by cadherins. The amount of cadherins expressed on the cell surface can be varied by using different clones of genetically engineered cells. This corresponds to modulating the cell–cell adhesion energy $W_{cc}$.

Adhesion to the substrate is achieved through integrins that bind to fibronectin deposited on the substrate. By coating the substrate with a variable amount of fibronectin (mixed with polyethylene glycol (PEG) to avoid non-specific interactions), the cell–substrate adhesion energy $W_{cs}$ is controlled. $W_{cs}$ can also be varied by tuning the stiffness of fibronectin-coated substrates.

As for the wetting of a surface by a drop (Section 4.4), we define the spreading parameter $S = W_{cs} - W_{cc}$. Here, this parameter can be varied. We can therefore quantitatively study the behavior (or more precisely the spreading) of a spheroid as a function of the sign of $S$ (Figure 10.30)[7]:

For $S < 0$, wetting is partial. With a substrate rich in PEG, there is a strong cohesion. Cell–cell adhesion is greater than cell–substrate adhesion. The aggregate forms a spherical cap, characterized by the contact angle $\theta_E$.

For $S > 0$, wetting is complete. The aggregate flattens, and a precursor film formed by a monolayer of cells spreads around the aggregate. Depending on the value of $W_{cc}$, this film is cohesive in a liquid state if $W_{cc}$ is large, whereas the precursor film is in a gaseous state for a low $W_{cc}$ value, and cells escape one by one from the aggregate.

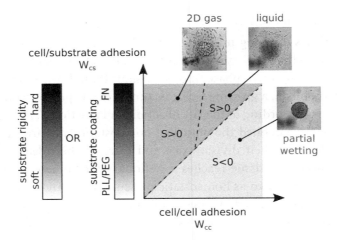

FIGURE 10.30 Wetting of a cellular aggregate on a surface as a function of the cell–cell $W_{cc}$ and cell–substrate $W_{cs}$ interaction energies. Depending on the sign of the spreading parameter $S = W_{cs} - W_{cc}$, partial or complete wetting situations with a cohesive or gaseous precursor film are observed on these images taken at long times. FN indicates a surface treatment with fibronectin, which induces strong cell adhesion. PEG-PLL is a copolymer that adsorbs on glass and greatly reduces cell adhesion. (Adapted from [7].)

### 10.5.3 Multicellular Aggregates as Foams

An alternative physical approach used to describe cellular aggregates and biological tissues has been to consider them as foams (Section 6.3). Indeed, the hexagonal shape of the cells in tissues is reminiscent of bubbles in foams.

This analogy, studied by F. Graner and his colleagues [8], is supported by the fact that a spheroid strongly squeezed between two plates does not completely recover its initial spherical shape once the load is withdrawn. This apparent plasticity, also present in embryo morphogenesis, results from the existence of topological rearrangements between adjacent cells that are identical to those observed in foams (Figure 10.31).

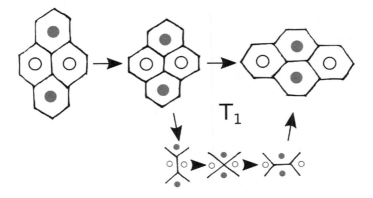

FIGURE 10.31 T1 topological transition observed in biological tissues and in foams.

## 10.5.4 Specific Material Properties of Multicellular Aggregates

One of the specificities of living matter compared to inert soft matter is that some cells divide and others die. The tissue, seen as a material, therefore spontaneously increases in volume when it is in the growth phase (more divisions than deaths). As a result, placed in the context of an organism, a growing tissue or tumor exerts pressure on the surrounding tissues, and vice versa according to the law of action and reaction. J. Prost, J.F. Joanny, and their colleagues [9] have defined the notion of homeostatic pressure. A pressure exerted on cells tends to decrease their division rate and increase their death rate. When these two rates are balanced, the tissue reaches a stationary (or homeostatic) state. This state is reached at a pressure known as homeostatic pressure. In a cancer context, tumor cells, which are generally characterized by uncontrolled proliferation, can therefore be thought of as giving rise to a high homeostatic pressure, likely to expand and potentially to cause metastasis.

## References

1. M.S. Steinberg, *Science*, 141, 401–408 (1963).
2. R.A. Foty, C.M. Pfleger, G. Forgacs, M.S. Steinberg, *Development*, 122, 1611–1620 (1996).
3. N. Bufi, P. Durand-Smet, A. Asnacios, *Methods Cell Biol.*, 125, 187–209 (2015).
4. Philip L. Townes, J. Holtfreter, *J. Exp. Zool.*, 128, 53–120 (1955).
5. M.L. Manning, R.A. Foty, M.S. Steinberg, E.M. Schoetz, *Proc. Natl Acad. Sci. USA*, 107, 12517–12522 (2010).
6. K. Guevorkian et al., *Phys. Rev. Lett.*, 104, 218101 (2010).
7. D. Gonzalez-Rodriguez et al., *Science*, 338, 910–917 (2012).
8. P. Marmottant et al., *Proc. Natl Acad. Sci. USA*, 106, 17271–17275 (2009).
9. M. Basan et al., *HFSP J.* 3, 265–272 (2009).

## 10.6 ENTANGLED ACTIVE MATTER

### 10.6.1 Definition

Biological tissues (Section 10.5) and swarms of ants belong to the vast domain of the "active matter," a newly recognized class of out-of-equilibrium materials, composed of many units that individually consume energy and generate collective movements. Active systems cover a wide range of length scales, from molecular motors to individual cells, tissues and organisms, and animal groups.

The components of a group can be free, such as fish in a pond (Figure 10.32A), or bound to each other by adhesive patches, such as ants in a swarm (Figure 10.32B). The latter situation is an example of "entangled active matter," a concept that has recently emerged to provide a unified description of their behaviors [1].

Entangled objects can be long polymer chains or piles of staples, which form transient networks due to topological constraints. The focus here is on active entangled systems formed by self-propelled adhesive entities, which cling by transient bonds. We illustrate this category with the example of ants attached to each other via adhesive patches and hairy legs (Figure 10.32B).

FIGURE 10.32 (A) Active matter: school of fish (Copyright Shutterstock). (B) Entangled active matter: ant raft. (Courtesy of David L. Hu.)

These transient bonds lead to a viscoelastic behavior. When squeezed between two plates, cell aggregates and ants are elastic under compression and relax like a rubber within short periods of time. In the longer term, they are viscous and flow like honey. This behavior reflects the high noise produced by living cells or ants moving in a swarm, much higher than thermal agitation, which allows them to relax the stresses exerted onto them.

## 10.6.2 Self-Adhesive Ants

The way ants organize and share work is fascinating. Everything is organized around the queen, the only one capable of giving life. After fertilization by winged males, who immediately succumb after mating, the queen lays up to several hundred thousand eggs that will give larvae. Nurses take care of baby ants while workers look for food and warriors protect the ant farm. Researchers are trying to understand how specialization is established in ant society: for example, if workers are eliminated, will nurses be able to fetch food and evolve as workers? But here, we will look at the ants with the eye of a physicist.

The fire ant, which originates from the rain forest in Brazil, has invaded North America where it causes multiple damages, especially in electrical installations. But their specialty is to resist flooding by making rafts that will allow the ant farm to survive for several months and protect the queen.

Thanks to adhesive patches and its hairy legs which entangle, the fire ant is able to build towers or climb on a Teflon tube to which it cannot adhere (due to weak Van der Waals interactions). The profile of the tower allows to calculate the number of ants carried by each one. It can also build rafts: the submerged ants, which support the structure, have a small air bubble that allows them to breathe for a while before they exchange places with the ants in the raft. We were all surprised to see many colonies of red ants floating as rafts of several kilometers in length in the flooded streets of Houston, USA, after Hurricane Harvey!

They are studied at Georgia Tech, USA, by David Hu [1], who wants to make self-assembled miniature robots that can work under extreme conditions.

## 10.6.3 Mechanical Properties of Balls of Ants

It is very easy to make ant balls: throw ants into water; because they float, collect them with a bowl. Put them in a glass and shake it. Ants stick together because of their adhesive patches and form a ball of about one centimeter.

By squeezing the ball, one can measure the elastic modulus $E$ and demonstrate the viscoelastic behavior of the ant ball: when quickly compressed and released, the ball becomes spherical again. By maintaining the compression for a while before release, the ball remains flattened.

Viscosity could be measured by dropping a coin into a container filled with ants. A more precise method is performed with a rheometer equipped with two Velcro strips to prevent ants from slipping off the walls (Figure 10.33). The density of ants can be varied, which modifies the properties of the transient gel formed by ants. The viscosity of ant swarms is about $10^6$ cP, one million times more viscous than water but 100 times less viscous than biological tissues. And remarkably, their density is about 0.2 g/mL, five times lower than water.

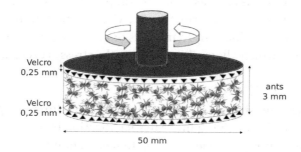

FIGURE 10.33    Rheometer to measure ants' viscosity as a function of shear and density. (Adapted from D. Hu.)

## 10.6.4 Wetting of Balls of Ants

It is also remarkable to note that living ants behave like liquid droplets.

Figure 10.34 shows two balls of ants that merge. The merging time $T_m$ is given by the scaling law $V^* T_m = R$, where $V^*$ is the capillary velocity ($V^* = \gamma \eta^{-1} \sim 10^{-8}$ ms$^{-1}$ for cells and $\sim 10^{-3}$ ms$^{-1}$ for ants). The fusion takes about one hour.

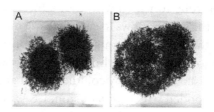

FIGURE 10.34    Merging of ant swarms. (Courtesy of David L. Hu.)

In the case of ant balls spreading on a solid substrate, a precursor film in a gaseous state is observed: ants escape from the aggregate towards freedom one by one. In the case of spreading on liquid, the precursor film is in a liquid state: ants must remain together to avoid drowning. On average, the ants' film is made up of two and a half layers, which allows the submerged ants to exchange places with the ants in the air.

### 10.6.5 Applications

Studies on red ants have applications in agriculture and robotics. Fire ants owe their name to the pain of their poisonous bite. They are an invasive species in the United States and are considered a harmful organism. They cause losses of more than $1 billion a year due to crop damage, sting injuries, and destruction of property.

The cooperativeness of ants has largely inspired modular robotics, i.e. the design and construction of robots capable of connecting together and forming self-assembled active particles. Indeed, as technology advances, robots are built smaller and smaller and look like ants in their abilities. Modular robotics has potential applications in the exploration of difficult terrain, such as in nuclear power plants or in extraterrestrial exploration.

Currently, modular robots have a number of limitations, which makes ant studies useful for inspiring new ideas for improving them. Indeed, modular robots can reliably connect to each other in a relatively small number of 2 to 1,000 individuals, which is small compared to the several hundred thousand ants that make up a colony.

### Reference

1. D.L. Hu et al.,, *Eur. Phys. J.*, 225, 629–649 (2016).

# CHAPTER 11

# Conclusion

THE AIM OF THIS book is to give an overview of the main fields of soft matter physics. It is also an opportunity to pay tribute to Pierre-Gilles de Gennes (Figure 11.1), who created this discipline. The authors attempted to write the different chapters by following de Gennes' style of teaching and doing research, i.e. by using scaling arguments to make complex calculations accessible to a wide audience and by being guided by curiosity. We explore many phenomena, technological achievements, and biological processes to which soft matter physics concepts may be applied. The different chapters are illustrated by some de Gennes' handmade slides and drawings. Drawing was his passion, and he got a price of the city of Paris for his synthetic style! He represents polymers in good or bad solvent in a very unique and artistic view (Figure 11.2).

Let us then finish this book with his views, by quoting the concluding remarks of a book chapter "Soft matter: birth and growth of concepts" [1].

FIGURE 11.1    Pierre-Gilles de Gennes in 1991 when he was awarded the Nobel Prize.

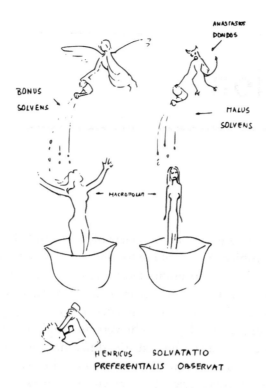

FIGURE 11.2 Cartoon made by P.-G. de Gennes for the 70th birthday celebration of the distinguished polymer experimentalist, Henri C. Benoit.

*Since the days of silex and potteries. hard matter and soft matter have coexisted. In this 20th century. the first half saw a cascade of "scientific supernovaes": relativity, quantum mechanics, microscopic physics, which related more directly to "hard" systems. The second part of the century showed one very unsuspicious supernova (molecular biology). Another one may be ready to explode (brain function). Some parts of our sky remain dark (e.g. fully developed turbulence). There is also a high noise level in all directions of observation (claimed discoveries which collapse, unrealistic simulations of natural phenomena…).*

*In this stormy world, soft matter, as we defined it, appears as a very small sector. But it represents the science of everyday life and, as such, it should take an increasing share in our educational system: up to the last century, most children lived in agricultural surroundings, and learned a lot -watching the birds, herding the sheep, repairing tools. Now this experience is lost: our school system ignores it and focuses on abstract principles. We need an education on simple things.*

*Insects have (temporarily) lost their grasp on the earth, because of their clumsy, hard, crust. Man has won, when its soft hand allowed him to make tools and to think. Soft is beautiful.*

## Reference

1. *Twentieth Century Physics*, eds by L.M. Brown, A. Pais and Sir B. Pippard. IOP Publishing Ltd, AIP Press Inc., 1995.

# Index